Laboratory Protocols for Conditional Gene Targeting

Laboratory Protocols for Conditional Gene Targeting

Raul M. Torres

Ralf Kühn

Institute for Genetics
University of Cologne

Oxford   New York   Tokyo
OXFORD UNIVERSITY PRESS
1997

Oxford University Press, Great Clarendon Street, Oxford OX2 6DP

Oxford New York

Athens Auckland Bangkok Bogota Bombay Buenos Aires
Calcutta Cape Town Dar es Salaam Delhi Florence Hong Kong
Istanbul Karachi Kuala Lumpur Madras Madrid Melbourne
Mexico City Nairobi Paris Singapore Taipei Tokyo Toronto Warsaw

and associated companies in
Berlin Ibadan

Oxford is a trade mark of Oxford University Press

Published in the United States by
Oxford University Press Inc., New York

A catalogue record for this book is available from the British Library

Library of Congress Cataloging in Publication Data
(Data available)
ISBN 0 19 963677 X

Typeset by the author
Printed in Great Britain by
Information Press Ltd., Eynsham, Oxon.

Foreword

This book grew out of Raul Torres' initiative to put together our lab protocols on gene targeting, so that newcomers would have a solid basis for starting their work. This became mandatory in particular when, due to Hua Gu's efforts, conditional gene targeting became feasible and many students and postdocs in the lab began to work in this area. When I saw Raul's text, which came as a surprise, and noticed the lively and encouraging style in which he was able to write, I immediately suggested he consider publication. At this point Ralf Kühn, member of our initial gene targeting team and the first to achieve inducible gene targeting, joined the venture, and the focus of the book became conditional gene targeting mediated by the Cre/loxP system.

I am delighted that the book is now ready for publication and think that it comes at the right time. Conditional gene targeting is just getting off the ground, and, offering ultimately to introduce mutations into any predetermined gene of the mouse in any cell type at any time of development, has a broad perspective. For all those who want to use this new tool the book should be useful in daily life. Let us hope that this approach will lead to novel insights into how vertebrate genes control biological functions in vivo!

January 1997, Klaus Rajewsky

"...anyone who leaves behind him a written manual, and likewise anyone who receives it, in the belief that such writing will be clear and certain, must be exceedingly simple-minded..."

Plato, Phaedrus

Preface

Embryonic stem (ES) cells retain the pluripotency of early embryonic-derived cells and, as such, have the potential to direct the development of a mouse. Coupled with the ability to identify homologous recombination events within these cells, the ability to manipulate ES cells has revolutionized the way we are able to approach biological questions. For this reason, the work of the many embryologists, cell biologists, and biochemists who pioneered these techniques deserve special recognition in making gene targeting the powerful technique that it has become today.

Indeed, from the first report of gene targeting in murine ES cells by Thomas and Capecchi (1), an impressive progress in the analysis of gene function in the mouse has been achieved through numerous mutants (2-4). As a result, the popularity of gene targeting is still increasing as many investigators turn to this technique to address their particular biological question (4-8). There has been, however, little variation on the strategy of how gene targeting vectors are constructed and on the type of genetic alterations generated by them.

A minimal generic targeting vector used for gene inactivation or modification will generally possess a region of homologous sequence juxtaposed to a selectable marker. Variations on this basic structure depend on whether genetic material is to be inserted or replaced during the homologous recombination event (1, 5, 6, 9). The end result with either of these strategies is usually the presence of the selectable marker at the targeted locus (simply inserted or replacing endogenous genetic material), which inactivates gene function. Several sophisticated variations of this general strategy, referred to as 'hit-and-run' and 'double replacement' (see below) have been established which aim to introduce subtle mutations or minimize the perturbations generated at the locus of interest.

To achieve the same purpose, in Klaus Rajewsky's unit at the Institute for Genetics in Cologne, Hua Gu pioneered an alternative strategy which exploits the bacteriophage Cre/loxP recombination system in combination with existing ES cell technology. In addition to the generation of subtle mutations, this system allows for a number of other genotypic options in ES cells by strategically incorporating loxP sites into the targeting vector, and subsequent expression of Cre recombinase, *after* homologous recombinants have been identified. Furthermore, when Cre is expressed in transgenic mice (and crossed to a loxP-containing target gene) the desired gene modification can be made conditional based on the expression properties of the promoter region used to drive Cre expression. The usefulness of this technology has already become apparent as a valid approach for the analysis of in vivo gene function. The numerous applications of the Cre/loxP recombination system significantly expands the potential of gene targeting methods (10-13) and should further stimulate genetic analysis in the mouse.

This book is not necessarily intended to be a replacement for the existing useful guides on embryonic stem cells and gene targeting (14-16). Rather, with this book we have attempted to summarize the capabilities of, and our experience with, the Cre/loxP recombination system as used together with gene targeting. The protocols found in this manual are a collection of bench-tested protocols which we have adopted for gene targeting in our lab and with particular emphasis on using the Cre/loxP recombination system. Our hope is that the many newcomers to gene targeting, as well as the experienced researchers in this field, will find this advice valuable.

Raul M. Torres

Basel Institute for Immunology
Basel, Switzerland

Ralf Kühn

Institute for Genetics
University of Cologne
Cologne, Germany

April 1997

Acknowledgements

We would like to express our thanks to Klaus Rajewsky under whose auspices this technology, and this manual describing it, was established and developed in Cologne. We also acknowledge all present and past colleagues from the Rajewsky lab for discussion and sharing unpublished results, especially: Marat Alimzhanov, Uli Betz, Björn Clausen, Irmgard Förster, Hua Gu, Steffen Jung, Daisuke Kitamura, Werner Müller, Roberta Pelanda, Klaus Pfeffer, Robert Rickert, Jürgen Roes, Frieder Schwenk, Eiichiro Sonoda, Takayuki Sumida, Shinsuke Taki, Alexander Tarakhovsky, Nobuyaki Yoshida, and Yongrui Zou.

Table of contents

Abbreviations

bp	Basepairs
DMEM	Dulbecco's modified Eagle's medium
DMSO	Dimethylfulfoxide
DNase	Deoxyribonuclease
EF	Embryonic fibroblasts
ES	Embryonic stem (cells)
FCS	Fetal calf serum
G418	Geneticin
GANC	Gancyclovir
HEPES	N-2-hydroxyethylpiperazine-N'-2-ethane sulphonic acid
HPRT	Hypoxanthine-guanine phosphoribosyl transferase
i.p.	Intraperitoneal
IU	International units
kb	Kilobasepairs
LIF	Leukemia inhibitory factor
mmc	Mitomycin c
MT-PBS	Mouse isotonicity PBS
neo	Neomycin resistance gene
PBS	Phosphate-buffered saline
p.c.	Post coitus
PCR	Polymerase chain reaction
(HSV)-tk	(Herpes simplex virus) thymidine kinase
SDS	Sodium dodecyl sulfate

Section 1
The Cre/loxP recombination system and gene targeting in ES cells

Chapter 1
Gene targeting strategies

Gene targeting, defined as the introduction of site-specific modifications into the mouse genome by homologous recombination, is generally used for the production of mutant animals to study gene function in vivo. Since homologous recombination of foreign DNA with endogenous genomic sequences is a relatively infrequent event in mammalian cells, compared to its random integration, the only efficient gene targeting method presently established utilizes pluripotent murine embryonic stem cell lines. Using these cells, the selection of rare, homologous recombinant ES cell clones in vitro can be accomplished. When such genetically modified ES cells are introduced into a preimplantation embryo they can contribute, even after extensive in vitro manipulation, to all cell lineages (including germ cells) of the resulting chimeric animal. The breeding of germline chimeras, which transmit an ES cell derived mutant chromosome(s) to their progeny, allows the establishment of an animal heterozygous for your genetic alteration and, importantly, by further breeding a homozygous mutant mouse strain. The necessity of chimera production may be bypassed in the future when spermatogonial stem cell lines become available (17, 18).

The production of a gene targeted mutant strain is a laborious, technically demanding effort which often takes more than a year to complete, and often even longer if technical problems are encountered. Therefore, it must be emphasized that a great deal of care and foresight should be given to this endeavor to ensure that the strategy chosen for the targeted modification of a gene will later also fulfill your experimental needs when a mutant mouse strain has been generated from targeted ES cells.

As a substrate for homologous recombination, vectors of the replacement type are most frequently used for gene targeting in ES cells and usually to simply inactivate gene function. A typical replacement vector consists of two regions of DNA (4-10 kb in total) homologous to the genomic target locus and which are interrupted by a positive selection marker such as the bacterial aminoglycoside phosphotransferase (neo) gene which is selected for with G418. This marker is either inserted into the homology region or, alternatively, replaces genomic sequence located between the homology arms (Figure 1a). If homologous recombinant ES clones will be identified by PCR one of the vector arms must be kept relatively short (1-2 kb) to ensure efficient amplification. A thymidine kinase (tk) gene is often included at the end of the long homology arm of the vector and serves as an additional negative selection marker (using gancylovir) against ES clones which have randomly integrated the targeting vector. Thus, homologous recombinants can be enriched by both positive and negative selection (19). The positive selection marker is primarily used to enrich for the rare stably transfected ES cell clones (obtained at a frequency of about one in 10^4 cells by electroporation) but frequently also serves as a mutagen, if appropriately designed, by insertion of the marker gene into a coding exon or the replacement of coding exons of the target gene by the marker. In addition, a small, nonselectable (e.g. point) mutation can be introduced into the homology region of the vector which is cotransferred into the target locus together with the selection marker in a fraction of homologous recombination events. In any case, the end product of a targeting experiment using a replacement type vector which includes a positive selection marker is the presence of the selection marker gene in the targeted locus which cannot be further modified. Hence, this gene targeting strategy is only suitable for the generation of a nonfunctional (knock out) allele of a target gene but should not be used for the introduction of subtle mutations since the selectable marker may disturb the regulation and splicing of the modified gene (20, 21). However, this precaution does not exclude, even if the target gene should be simply inactivated, possible interference of the selectable marker with genes adjacent to the target gene or coded on the opposite strand (22-24). Moreover, when using the positive selection marker to inactivate gene function, careful consideration should be given to potential outcomes besides the desired gene inactivation. These would include alternative splicing of the affected exons (25) and/or the potential generation of dominant negative molecules.

To circumvent the problems associated with the presence of a selection marker in a targeted locus several two-step gene targeting strategies have been developed which allow the introduction of subtle, nonselectable mutations into, and the removal of the selectable marker from, the targeted gene. The 'hit-and-run' method (9, 26, 27) also termed 'in-out' (28), utilizes an insertion type targeting vector which is linearized for transfection within the homology region of the target gene. In the first targeting step ('hit'), homologous recombination leads to the complete integration of the vector into the target locus generating a partial duplication interrupted by plasmid sequences and selection markers (neo as positive and tk as negative marker; Figure 1b). In the example shown in Figure 1b a point mutation was introduced in exon 2 of the homologous sequence contained within the targeting vector. The duplication can be resolved by an *intra*chromosomal homologous recombination event or unequal sister chromatid exchange ('run') occurring spontaneously at a low rate in a population of targeted ES cells. Such (tk-negative) cells can be enriched from the tk-positive majority using selection (e.g. gancyclovir) which kill tk-expressing cells. If recombination occurs as depicted in Figure 1b the point mutation in exon 2 is incorporated into the restored target gene while all heterologous sequences are lost from the targeted locus. However, the intrachromosomal recombination event can also result in the restoration of the wildtype allele depending on the actual point of exchange. For a detailed account of hit and run targeting vector construction and identification of targeted clones see (27).

a) Gene inactivation

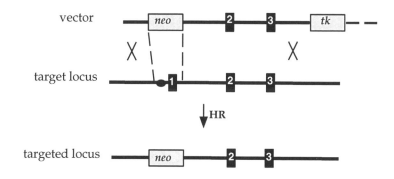

b) Hit and run

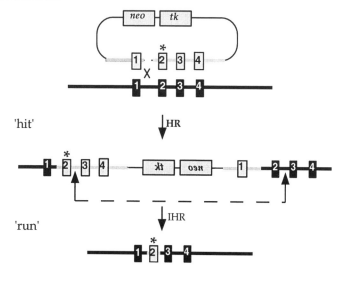

c) double replacement

1. targeting vector

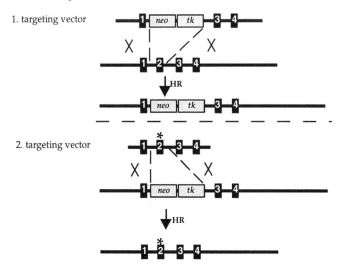

d) coelectroporation

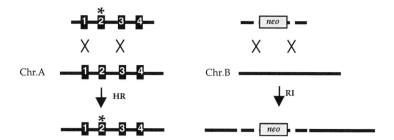

Figure 1. Conventional gene targeting strategies. Cre/loxP based gene targeting is depicted in Figures 4-12. Symbols are as indicated in Figure 4. Extrachromosomal recombination products are not shown. **(a)** Simple gene inactivation with a replacement type vector. The promoter and first exon (filled rectangle) of the target gene are exchanged by a neomycin resistance gene (neo) by homologous recombination (HR) with the linearized targeting vector. Crossover points are depicted by an X. The tk gene and plasmid sequences (thick stippled line) are lost during homologous recombination. **(b)** Introduction of a point mutation by the 'hit-and-run' method. The integration type targeting vector, containing neo and tk genes within the plasmid backbone (thin line), is linearized within the homology region of the target gene (shaded lines and rectangles) between exons 1 and 2 (rectangles). A point mutation (asterisk) is introduced in exon 2 of the targeting vector. In the first (hit) step the target locus (solid black line and rectangles) is

partially duplicated by the complete integration of the targeting vector between exons 1 and 2. The point mutation is located in most cases in the 5' duplicate but transfer of the mutation to the 3' duplicate due to branch migration or loss of the mutation as a result of heteroduplex repair may occur (not shown). After identification of homologous recombinant ES clones, tk-negative ES cells, which have undergone an intrachromosomal homologous recombination (IHR, arrows) event or uneven sister chromatid exchange within the duplicated locus, are enriched by gancyclovir selection. **(c)** Introduction of a point mutation using the double replacement method. In the first targeting step (1. vector) exon 2 of the target gene is replaced by neo and tk genes. Homologous recombinant ES cells are transfected with a second targeting vector containing a point mutation in exon 2 (asterisk) of the target sequence but no selectable marker. Tk-negative ES cells, which exchanged the neo and tk genes by homologous recombination with the second targeting vector, are enriched by gancyclovir selection. **(d)** Coelectroporation of a targeting vector with a point mutation (asterisk) in exon 2 of the target sequence and a second, independent vector containing the neo gene as selectable marker. Among the G418 resistant ES cell population, which randomly integrated (RI) the selection marker (e.g. on chromosome B), clones must be identified which homologously recombine the target gene (on chromosome A) with the targeting vector.

In contrast to hit-and-run, the 'double replacement' or 'tag-and-exchange' method (29-31) involves two sequential gene targeting steps using two independent targeting vectors for the introduction of a nonselectable mutation into a target gene. In the first step neo and tk genes are inserted into the target locus using a replacement type vector (Figure 1c). In the second targeting step homologous recombinant ES clones are transfected with a replacement type vector which replaces the neo and tk genes in the targeted locus and introduces a nonselectable (e.g. point) mutation (Figure 1c) into the restored target gene. As in the case of the hit-and-run method, the second recombinant ES clones are enriched by their tk-negative phenotype using gancyclovir selection. More efficiently, HPRT can be used as negative selection marker in $hprt^-$ ES cells (29). If many replacements are desired at the same target locus double replacement is superior to the hit-and-run method since the first replacement needs to be performed only once and all further replacements can be done in a single step. For both methods, however, the efficiency of negative selection in the second targeting step is critical for success. This may be overcome by using a modification of the double replacement ('plug and socket'; (32) which takes advantage of, although is not restricted to, HPRT-deficient ES cells for the positive selection of the second targeting event.

The coelectroporation of a targeting vector harboring a nonselectable mutation together with a second, independent vector containing a selection marker (neo) has been tested at the *hprt* locus as an alternative, third method for the introduction of subtle mutations (33, 34). Among the G418-resistant ES cell colonies obtained from transfection, which randomly integrated the vector containing the selection marker, clones must be identified which have simultaneously incorporated the subtle mutation into the target locus by homologous recombination with the second transfected vector (Figure 1d). The frequency at which the desired homologous recombinant clones with additional random integration of the selectable marker were found, however, was 10-20-fold lower compared to the use of a conventional replacement vector. Thus, coelectroporation does not appear to be beneficial for most gene targeting applications, although may be useful for targeting loci which are anticipated to suppress the expression of selectable marker genes.

The use of the recombinase systems, and in particular Cre/loxP (Chapter 2), is the latest addition to the technical improvements in gene targeting. Its various applications, which can be all accomplished using replacement type vectors, are discussed in Chapters 4-11 of this book. Like the hit-and-run and double replacement methods, Cre/loxP based gene targeting allows the introduction of subtle mutations into target loci with the subsequent removal of selection marker genes. The former methods, however, have the potential advantage that no heterologous sequence remains in the targeted locus whereas a single (34 bp) loxP site remains after Cre/loxP mediated gene targeting. Furthermore, double replacement may also be of advantage if many replacements should be made in the same target locus. On the other hand, the Cre/loxP system allows, apart from the introduction of subtle mutations, for a number of other genotypic options in ES cells by the strategic incorporation of loxP sites into the targeting vector (Figures 5-13). We regard the use of this recombination system as technically easier and more reproducible compared to the other two-step targeting methods. Moreover, use of the Cre/loxP system provides a greater versatility to your gene targeting experiment which, in addition to facilitating conventional gene inactivation by the creation of large deletions and/or the removal of the selection marker from the targeted locus, also affords an opportunity for conditional gene modification in vivo.

Clearly, choosing a suitable strategy is the first important thing you have to think about if you are planning a gene targeting experiment. If the goal is only gene inactivation, a simple replacement vector with the selection marker as mutagen (Figure 1a) might be the most straightforward option. In this classical form of gene targeting the ES cell genome is modified to derive a mutant in which the target gene is inactivated, from fertilization on, in all cells of the body throughout ontogeny (Figure 2a). This strategy may be most appropriate to test for the function of genes suspected to have a role during embryonic development or for new genes of unknown function since the whole animal can be screened for effects of the genetic modification. It is certainly also a suitable method to study the function of genes which are expressed in a regionally restricted pattern in adult mice.

In contrast to the classical method, conditional gene targeting can be defined as a gene modification which is restricted to certain cell types or developmental stages of a mouse (Figure 2b, 2c). The use of the Cre/loxP recombination system for conditional gene targeting requires the generation of both a mouse strain harboring a loxP-flanked segment of a target gene and of a second strain expressing Cre recombinase in specific cell types. A conditional mutant is generated by crossing these two strains such that the modification of the loxP-flanked target gene is restricted in a spatial and temporal manner according to the pattern of Cre expression in the particular strain used (see Figure 13 and Chapter 11 for details).

This strategy allows us to inactivate or modify genes in a cell type-specific manner either constitutively, from a certain developmental stage on (depending on the particular regulatory elements used for Cre expression) (Figure 2b), or from a given timepoint on, upon induction of Cre activity (Figure 2c). The regional specificity of conditional gene targeting can be used to test the function of widely expressed molecules in a particular cell lineage and to investigate gene function postnatally if conventional gene inactivation leads to a severe or lethal phenotype during embryonic development. Inducible gene targeting should be especially helpful to analyze gene function in adult animals since it allows gene modification after the normal establishment of adaptional responses (e.g. memory).

There is no definitive rule to decide whether, for a particular experiment, conventional or conditional gene targeting is more appropriate since this depends on your specific biological question and the peculiarities of the gene studied. In fact, the answer to this question may not be predictable in many cases but will require the performance of the actual experiment. However, following one of the strategies described in Chapter 9, both conventional and conditional gene targeting can be applied to a particular gene requiring only a single targeting vector and germline transmission of only one mutant allele.

Wildtype

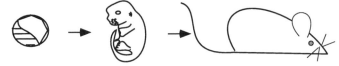

Conventional gene targeting

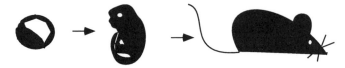

Cell type-specific gene targeting

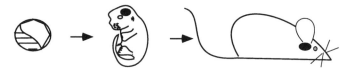

Inducible gene targeting

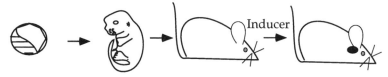

Figure 2. Comparison of conventional and conditional gene targeting strategies at different ontogenetic stages (blastocyst; embryo; adult). Filled, black regions indicate tissues/organs expressing the modification of a target gene whereas white regions symbolize its wildtype function. **(a)** Wildtype mouse. **(b)** Conventional mutant expressing a mutation in all tissues throughout life. **(c)** Constitutive, cell type-specific gene targeting. In this example a promoter region is used for Cre expression which becomes active in a certain cell type and region of the brain (black oval) during embryonic development. **(d)** Inducible, cell type-specific gene targeting. The activity of Cre recombinase is switched on in certain cell types upon administration of an inducer; in the example shown Cre should be inducible in a certain cell type and region of the brain.

Chapter 2
The Cre/loxP recombination system

The P1 bacteriophage Cre recombinase recognizes and mediates site-specific recombination between 34 bp sequences referred to as loxP (locus of crossover (x) in P1) (35). The loxP sequence consists of two 13 bp inverted repeats interrupted by an 8 bp nonpalindromic sequence which dictates the orientation of the overall sequence (Figure 3a) (36, 37). When two loxP sites are placed in the same orientation on a linear DNA molecule, a Cre-mediated intramolecular recombination event results in the excision of the loxP-flanked, or 'floxed', sequence as a circular molecule with one loxP site left on each reaction product (Figure 3b). The reverse reaction, an intermolecular recombination event, will result in the integration of the circular DNA molecule into the linear molecule, each possessing one loxP site. Alternatively, an intermolecular recombination event between two linear molecules as substrates (again, each possessing one loxP site) results in the reciprocal exchange of the regions flanking the loxP sites (Figure 3c). If the two loxP sites are placed in opposing orientation the floxed sequence will be inverted (Figure 3d) (38-40).

Mutant loxP sites with single nucleotide exchanges in the spacer region can still be recombined efficiently with each other but not with wildtype sites, whereas a symmetric spacer region leads to an equal frequency of excision and inversion (41). This feature appears to add further versatility to the use of the Cre/loxP system allowing Cre to recombine different sequences independently without worry of interchromosomal recombination events. By placing two different point mutations into different halves of a pair of lox sites the equilibrium of the Cre-mediated reaction can be shifted towards the products since the double mutant and the wildtype lox sites, generated by recombination, are less efficiently recombined than the mutant substrates (42). This property can be used to favor Cre-mediated integration by preventing the reexcision of a loxP-flanked DNA segment.

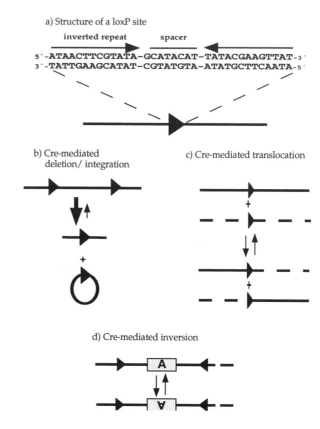

a) Structure of a loxP site

inverted repeat   spacer

5`-ATAACTTCGTATA-GCATACAT-TATACGAAGTTAT-3`
3`-TATTGAAGCATAT-CGTATGTA-ATATGCTTCAATA-5`

b) Cre-mediated
deletion/ integration

c) Cre-mediated translocation

+

+

+

d) Cre-mediated inversion

A

Ɐ

Figure 3. LoxP site structure and products of intra- and intermolecular Cre-mediated recombination between two loxP sites. **(a)** A 34 bp loxP site consisting of two 13 bp inverted repeats and an asymmetric 8 bp spacer region. The spacer region defines the orientation of the loxP site represented as a filled triangle. **(b)** Cre-mediated recombination between two directly repeated loxP sites on a linear DNA molecule leads to the excision and circularization of the loxP flanked DNA segment. One loxP site remains on each of the reaction products. In the reverse reaction a loxP containing circle is integrated into a linear DNA molecule. **(c)** Cre-mediated intermolecular recombination of two linear DNA molecules, each containing one loxP site. The regions flanking the loxP sites are reciprocally exchanged between the reaction partners as a result of recombination. **(d)** Cre-mediated inversion of a DNA segment flanked by two loxP sites in opposite orientation.

While the exact mechanism of Cre-mediated recombination is not entirely worked out, a few details of the reaction are known and some of the relevant observations are included here. During Cre-mediated recombination, which proceeds through a Holliday intermediate, tyrosine residue 324 forms a phosphodiester with the 3´-PO4 group of one of the strands of the parental DNA molecule which is then transferred to the 5´-OH group of the recipient strand (43, 44). Each of the two repeats within a loxP site is bound by one molecule of Cre (45) and a staggered 6 bp cut is generated. Two directly repeated loxP sites must be at least 82 bp apart to be recombined (46). In the life-cycle of the P1 phage Cre recombinase (causes recombination) serves to ensure the segregation of the cyclic unit-copy P1 plasmid prophage and the cyclization of the linear viral genome after infection of bacteria (47, 48).

The Cre recombinase is a 38 kDa protein classified as a member of the λ integrase superfamily of site specific recombinases (49, 50) and belongs, like the yeast derived FLP and R recombinases (51, 52), to the few members of this family which do not require cofactors or accessory proteins for recombination (44, 53, 54). It is this feature which makes the Cre/loxP system a useful tool for genetic engineering and it has been used successfully for the genetic manipulation of (trans)genes in mammalian cells (40, 55-58), mice (59-63), yeast (64), bacteria (38, 65) and plants (66-68). For the 1029 bp coding sequence of Cre we refer you to Sternberg (1986) and the EMBL database accession number X03453. Presently, the Cre recombinase is routinely used in ES cells and transgenic mice with high recombination efficiencies whereas the reported use of the FLP/FRT system in ES cells and transgenic mice has not been as efficient (20, 69, 70). Recently, Stewart and colleagues (71) have demonstrated that the yeast FLP recombinase operates optimally at 30°C, whereas the bacteriophage Cre recombinase has an optimum temperature of 37°C. This discrepancy in temperatures may explain the inefficiency of the FLP/FRT recombination system in mammalian cells. In addition, a point mutation has been identified in a commonly used mammalian FLP expression vector which even increases the thermolability of FLP and results in decreased recombination efficiency (71). Thus, for the in vivo manipulation of genes in mice, Cre/loxP appears to be clearly the recombination system of choice. Nevertheless, FLP-mediated FRT-specific recombination has been used for the removal of an FRT-flanked selection marker in ES cells (20, 69) and for gene manipulation in other mammalian cells (72, 73). FLP is also used successfully in yeast (51), Drosophila (74, 75) and plant cells (76, 77). Finally, the R recombinase has been used in yeast and plant, but not yet in mammalian cells (54).

Chapter 3
Gene targeting vectors - considerations and use of loxP

In general, the experimental outline of Cre/loxP mediated gene targeting is accomplished in two steps. The first is the incorporation of loxP sites into the genome of ES cells via a corresponding targeting vector and the identification of these homologous recombinant clones. The final genetic modification is obtained by a Cre-mediated recombination event, either stably in cultured ES cells (and the germline of mice derived from these ES cells) or in a conditional manner, in mice by the expression of Cre as a transgene. Because the potential of each targeting construct is greatly expanded with the use of this system, loxP sites should be generally considered for all targeting vectors even if the initial goal is simply to inactivate gene function.

The construction of a targeting vector can be arguably the most tedious aspect of gene targeting. A great deal of forethought should be invested into the vector design since homologous recombinants are generated only after a great deal of energy has been expended - both in the molecular biology to make the construct and the tissue culture to isolate recombinants. This should perhaps be emphasized even more so when using Cre/loxP in your targeting construct. Simple gene inactivation is usually straightforward with enough knowledge about the particular gene in question. Cre/loxP, however, can increase the options with each construct and a careful placement of an additional loxP site(s) within the construct (or in a separate targeting event) may facilitate further strategies in which to alter the genome in vitro or in vivo. On more than one occasion homologous recombinants have been identified when the realization that a 'better' construct could have easily been generated allowing yet another approach to your question.

Obviously each gene targeting experiment will have its own nuances, requirements, and limitations because of a number of variables and the same is true for the generation of each targeting construct. How much genomic DNA is available, how well the locus is characterized, what convenient restriction site(s) exist, etc.; questions which must be answered and improvised with each targeting construct. What follows here are some general issues to consider when generating a replacement type vector for gene targeting and, in particular, when using the Cre/loxP recombination system.

Optimally, the source of genomic DNA utilized in the targeting construct should be from the same strain as that of the target ES cells to avoid a reduced targeting frequency due to sequence polymorphism (78, 79). There have been, however, as many successful homologous recombinants using non-isogenic DNA as with isogenic DNA in our experience and there have been several examples of using the same construct with isogenic and nonisogenic ES cells with little difference in homologous recombination frequencies. It is always a good idea, however, to first examine the specific locus you are interested in for restriction fragment polymorphism if using nonisogenic DNA since polymorphism at the locus would suggest a lower frequency of homologous recombination.

The longer the region of homology the higher the frequency of homologous recombination (within practical limits) (1, 79), however, because we generally use replacement-type vectors and screen for homologous recombinants by PCR (or always wish to have the option), virtually all our targeting constructs possess a short and long arm of homology which flank the selectable marker gene. On practical terms, by 'short' arm we generally consider constructs $\geq$ 0.5 kb but short enough to be reliably amplified (< 1.5-2 kb) and detected (on ethidium bromide stained gels) with PCR. A 'long' arm of homology is generally greater than 4 kb with an upper limit dependent on a number of variables such as available sequence, feasibility in cloning, etc. These lengths are very general and are certainly not definitive, i.e., homologous recombinants have been generated using constructs which vary widely with respect to these criteria. If you have decided not to use PCR for screening ES colonies the homologous DNA should be more equally distributed among the two vector arms as aberrant recombination products are found more frequently using short vector arms (80). The placement of loxP sites within the homology regions, when Cre/loxP is to be used, depends on the nature of your specific question and will be discussed in the following sections. As a precautionary measure it would be wise to verify the integrity of the loxP sites in your vector before transfecting ES cells. This is perhaps most easily accomplished by using the targeting vector to transform a bacterial strain harboring the Cre gene in the genome (81). After overnight growth of the transformed bacteria the purified vector can be examined for appropriate Cre-mediated recombination.

As a selectable marker we generally use the nonmutated version (82) of the bacterial aminoglycoside phosphotransferase (neo) gene controlled by the Herpes Simplex Virus (HSV)-thymidine kinase (tk) promoter/polyoma enhancer region from pMC1neopA (1), yielding about

1000 G418 resistant ES colonies per 10^7 cells electroporated. 'Stronger' expression cassettes which can be expressed in a larger number of genomic sites, as measured by a higher yield of resistant ES colonies, have been described (83). However, stronger neo expression does not necessarily equal better selection, i.e., if a 'weak' neo gene like MC1neopA confers G418 resistance when integrated at a targeted genomic locus, the use of a more potent neo expression unit may simply increase the background of random integrants and thereby reduce the frequency at which homologous recombinants are isolated. A targeting experiment would fail, on the other hand, if a weak neo gene is used which is not sufficiently expressed at the targeted locus to confer G418 resistance to targeted ES clones. Since the MC1neopA gene has been used at our institution successfully for the isolation of homologous recombinants for more than two dozen genes the frequency of loci inhibiting MC1neopA expression seems to be rather low. We have experienced one case in which targeted ES clones appeared selectively late after transfection due to their slow growth in G418 containing medium. If the selection marker should be removed from a modified locus after the initial targeting step a loxP-flanked version of the neo gene must be used for vector construction (see Chapter 4).

The ability to both positively select for stable transformants as well as negatively select against random integrants may enrich the number of homologous recombinants found among the screened colonies. Typically we use the neo gene for positive selection (culturing transfectants in the presence of G418) and, when used, negatively select by including the HSV-tk gene (MC1-TK; see (19)) and gancyclovir treatment. In our experience enrichment by negative selection is usually in the range of three- to ten-fold resulting often in a frequency of 1/10 - 1/40 of homologous recombinant ES clones as compared to random integrants. When also using Cre/loxP an additional question must be considered: where (and whether) to use the tk gene. When used to select against random integrants, the tk gene is placed just outside the long arm of homology however it can also be used to positively select clones in vitro which have undergone a Cre-mediated excision event when situated between two loxP sites. The question whether to use the tk gene for either of these purposes (or at all) depends on which of these steps, the initial targeting event or the Cre-mediated deletion, is considered as most laborious in terms of identifying ES clones with the desired modification (i.e., which event will be less likely to occur). As discussed in the following sections Cre-mediated deletion is highly efficient for most applications in gene targeting and usually needs no further enrichment by tk/gancyclovir selection. As a result, we usually place the tk gene at the end of the long homology region of a targeting vector to enrich for the homologous recombination event.

Since homologous recombination occurs more frequently using linear DNA (84, 85) targeting vectors should be linearized prior to transfection of

ES cells. It is therefore mandatory to include a unique restriction site in the targeting vector at the end of one of the homology regions. We prefer to introduce the tk gene at the end of the long homology arm and to linearize our constructs just outside the short arm of homology, including a minimal amount of plasmid sequence, such that the plasmid backbone is attached at the end of the tk gene. The presence of short heterologous sequences and plasmid backbone at the vector ends does not diminish targeting frequency (19, 79) and the presence of plasmid sequence may even protect the tk gene from inactivation when random integration occurs. A linearized replacement vector, to be used with PCR screening, is schematically shown in Figure 1a.

Another important issue to consider *before* starting construction of a targeting vector is that you must be able to identify easily, either by PCR or directly by Southern blotting, a few targeted ES colonies among an excess of random integrants in the first screen after ES cell transfection. When clones have been identified as putative homologous recombinants in the first round of screening the presence of the correct targeting event must be unambiguously confirmed by Southern blotting using probes inside (internal) and outside (external) of the targeting vector to exclude clones with aberrant recombination products (1, 86). Thus, it is necessary for vector construction to clarify in advance which genomic fragments could be used as probes for Southern hybridization, which restriction enzyme digests distinguish wild-type, and correctly and inappropriately targeted alleles and, when using Cre/loxP, reveal the presence of loxP sites and differentiate deleted from nondeleted alleles. Both PCR primers and genomic probes should be confirmed to function adequately before ES cells are transfected with the targeting vector. These aspects are discussed in more detail in Chapter 17.

The following chapters give an overview of the presently established applications of the Cre/loxP system for gene targeting and the configuration of the corresponding loxP containing vectors. This survey is, however, certainly not complete as additional applications will continue to be found in the future.

Chapter 4

Nonselectable modifications and removal of selection marker genes

One of the most straightforward purposes for using Cre/loxP is to remove a loxP-flanked (floxed) selectable marker after homologous recombinants have been identified in a gene targeting experiment. This is certainly important for any experiment designed to modify, but not abrogate, gene function by preventing possible interference of the selectable gene (and its own regulatory elements) on gene function and/or regulation (20, 21, 87). The range of modifications which can be introduced into the target gene is virtually unlimited ranging from alterations in the coding region, like point mutations or exon insertion (or deletion), to changes in regulatory elements such as the promoter/enhancer region of a gene. Thus, the design of a targeting vector will vary depending on the nature of your specific question. In some instances the genomic sequences which are to be deleted may simply be replaced by the selectable marker gene in the targeting vector; elements to be inserted can be directly linked to the selection marker creating a single region of nonhomology in the vector. In many cases, however, it will be more appropriate to insert the selection marker (without replacing genomic sequence) into a different position than the desired modification in the targeting vector to avoid the mutation being directly neighbored by the loxP site, which remains in the genome after Cre-mediated deletion of the floxed marker gene. This is an obvious requirement if the coding region of a gene should be altered but might be equally important for the modification of regulatory regions. However, if the resistance marker and a nonselectable mutation are separated by homologous sequence they are not always cotransferred into the targeted locus since the breakpoint of recombination may be located anywhere in between, resulting in a fraction of resistant, targeted ES cell clones which have integrated only the selection marker (as this is the basis for selection). Since the frequency of cotransfer declines with increasing distance between two heterologous sequences, a nonselectable mutation should probably not be more than 10 kb away from the selection marker to avoid the need for isolation and screening of a large number of homologous recombinant ES clones. As an example, in the tumor necrosis factor locus frequencies of cotransfer between the neo gene and a 40 bp heterologous sequence (loxP) were found to be 70% over a distance of 4 kb, 35% at a 7 kb distance and 2% at a 10 kb distance (M. Alimzhanov, unpublished results). These frequencies will also of course depend on the amount of homology found distal of the heterologous sequence from the selectable marker. Therefore it is important, when planning a targeting vector for the introduction of a nonselectable mutation, to define an easy strategy for the detection of that mutation (e.g. by the inclusion of a restriction site).

After homologous recombinant ES clones containing the nonselectable mutation have been identified the floxed selection marker can be removed by the use of Cre recombinase. Consequently, a single loxP sequence remains in the targeted locus after deletion. To avoid potential interference of this loxP element with the function of the targeted gene it might be sufficient in many cases to simply avoid the presence of loxP in obvious control elements like the promoter or enhancer region or close to splice sites within introns. A good choice for the position of a loxP site might be large introns or the region downstream of the poly A signal sequence of a gene. Moreover, in some cases it might be necessary to generate a control mouse strain from ES clones which contain only the remaining loxP site but not the desired modification in the targeted gene, especially if this mutation is expected to interfere with the regulation of the targeted gene (88).

A generic gene targeting vector for the introduction of a nonselectable mutation using Cre/loxP is configured like a replacement type vector and contains a selectable marker flanked by loxP sites. Figure 4a schematically demonstrates the structure of a vector used to insert a nonselectable (point) mutation in an exon and the selection marker in an intron of a gene. If an additional exon should be inserted into a genomic locus the exon region might be combined with the selection marker into a single region of nonhomology in the targeting vector (Figure 4b). At our institution this strategy was used, for example, to insert rearranged antibody gene segments (including promoter) into the immunoglobulin (Ig) heavy and light chain loci for the production of mice with a single antibody specificity (89, 90, and R. Pelanda, unpublished results). As a selectable marker for vector construction we routinely use a floxed neo gene cassette from plasmid pL2neo or pMMneoflox8 (see Appendix B). This cassette can be efficiently removed from a targeted locus by transient transfection of targeted ES cells with a Cre expression vector. In our hands, using the pIC-Cre or pgk-Cre plasmids (see Appendix B) for transfection, deletion of the floxed neo gene occurs at a frequency at least 50% of the efficiency of the transient transfection. Thus, if 10% of the cells from a neo^{flox}-containing ES clone would be transiently transfected with Cre-expression vector, 5 of 100 ES cell colonies derived from this transfection can be expected to have lost the neo gene by Cre-mediated

19

deletion. Such clones can be easily identified, without the need for molecular biology, by their sensitivity to G418 (see Chapter 21 for protocol). We find that it is generally not necessary to use a loxP-flanked tk-neo cassette to enrich, by gancyclovir selection, those clones which have deleted the cassette. Hence, for the simple deletion of the selection markers, the tk gene should probably not be used for the enrichment of deleted clones, but rather should be included instead (if at all) at one of the ends of the targeting vector to select for homologous recombinants in the first step of the gene targeting experiment.

At our institution the Cre/loxP strategy to remove a selection marker from a targeted locus has been applied to introduce a stop codon into the coding region of a gene (similar to Figure 4a), with the neo^{flox} cassette inserted downstream of the polyA signal sequence (91), and to insert rearranged immunoglobulin variable regions into the Ig loci for the production of mice with single antibody specificity (Figure 4b; ref. (90) and R. Pelanda unpublished results). The targeted insertion of a V region into the Ig heavy chain locus was first described by Taki et al. (89), however, the FRT-flanked neo gene used in that experiment could not be deleted in ES cells using the FLP recombinase. Furthermore, several rounds of gene targeting within one ES cell clone can be performed using a single floxed selection marker by its removal from the genome of homologous recombinants (92).

As an alternative to the Cre/loxP system the 'hit-and-run' method (Chapter 2) can also be used to remove a selection marker from a targeted locus. In our view this method is more laborious and complicated compared to the Cre/loxP system but has the advantage that no heterologous sequence remains in the targeted locus after the 'run' step, while, at minimum, a 34 bp loxP site is left after Cre-mediated deletion. Thus, the use of the 'hit-and-run' method might be considered for experiments which must absolutely exclude any potential influence of a short heterologous sequence on gene regulation.

a) Introduction of a point mutation

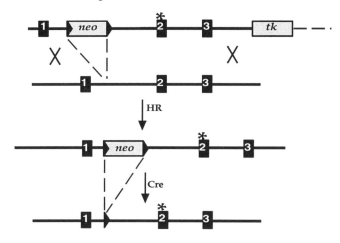

b) V gene insertion

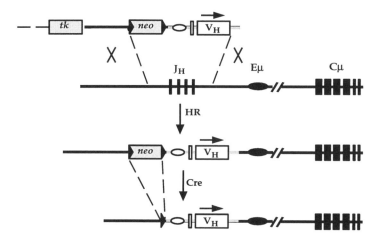

Figure 4. Removal of selection marker genes from targeted loci in ES cells using the Cre/loxP system. Targeting vectors are shown linearized at the end of the short homology arm, as used for electroporation. Thick stippled lines at the end of the tk gene represent the residual plasmid backbone. Continuos lines represent homologous sequences in targeting vectors, target and targeted loci. Heterologous genomic sequences are drawn as shaded lines. Exons are represented as numbered, filled rectangles, selection marker genes as stippled rectangles and loxP sites as filled triangles. The crossover points during homologous recombination (HR) are depicted by an X. Thin dashed lines connecting targeting vectors and genomic loci show the sites of insertion of selection markers and other heterologous sequences

into the genomic loci used for vector construction. Thin stippled lines connecting targeted loci before and after Cre-mediated recombination (Cre) show the deletion of loxP flanked sequences. Excised recombination products are not shown. **(a)** Introduction of a point mutation (asterisk) into an exon of a target gene in ES cells. A loxP flanked neo gene placed in the first intron of the target gene is introduced together with a point mutation in exon 2 into the target locus by homologous recombination. The floxed neo gene is excised by Cre-mediated recombination after transient expression of Cre recombinase in targeted ES cells. A single loxP site remains in the intron between exons 1 and 2. **(b)** Insertion of an immunoglobulin V_H gene into the Ig heavy chain locus in ES cells. In the targeting vector the V_H region (including a promoter) is combined with a loxP flanked neo gene into a single region of nonhomology replacing the J_H elements. After homologous recombination the V_H gene is controlled by its own promoter region (open ellipse) and the endogenous heavy chain enhancer (Eμ; filled ellipse) directing transcription (horizontal arrow) to the Cμ region. The loxP flanked neo gene can be removed from the targeted locus by transient Cre expression in ES cells.

Chapter 5
Large deletions

Small genomic deletions (1-2 kb) are often created by replacing genomic sequence with the selection marker gene included in a replacement type gene targeting vector. This strategy has been used to generate deletions up to 19 kb in length in the *hprt* locus with no decrease in the frequency of homologous recombination compared to smaller replacements in the same locus (93). The majority of recombination products, however, obtained in these latter experiments were found to have an aberrant structure with a rate of correctly targeted ES cell clones of 10 - 30%. In contrast to this single step method for generating genomic deletions, the Cre/loxP system requires that two loxP sites first be introduced by homologous recombination into a locus and the actual deletion is created in a second step by transient expression of Cre recombinase in targeted ES cell clones. Thus, the strategy for generating deletions of up to approximately 10 kb, using a single targeting vector, is the same as described in the section on the flox-and-delete approach (see Chapter 9). Compared to the conventional single step method, the Cre/loxP system has the advantage that virtually no unexpected recombination products are obtained in the Cre-mediated deletion step and that the selection marker is removed from the targeted locus. Moreover, unlike other methods, the use of the Cre/loxP system offers the unique possibility to generate very large deletions by the introduction of loxP sites into different chromosomal positions using two gene targeting vectors sequentially, each transferring one loxP site. This general strategy has proven successful in deleting DNA segments in ES cells ranging in size from 200 kb (94) to 3-4 cM (95). The potential applications of large deletions include the targeted inactivation of large genes, the removal of whole gene clusters from the genome, and genetic screening for recessive mutations.

To generate deletions up to about 10 kb a single targeting vector can be used. This vector must include a positive selection marker gene neighbored to a loxP site, or flanked by two loxP sites, and an isolated loxP site in the same orientation within the homology region of the vector

(Figure 5). After homologous recombinants are identified, the loxP-flanked genomic fragment and the selection marker are deleted by transient expression of Cre recombinase in ES cells (see Chapter 21). The use of a tk-neo-loxP cassette (e.g. from plasmid pGH1, see Appendix B) might be helpful to enrich for clones harboring the desired deletion but in our view would be optional because of the high efficiency of Cre-mediated deletion in the range of 1-10 kb. For the same reason, whether the selection marker cassette includes one loxP site or is flanked by loxP sites is irrelevant (if the deletion is the only objective). Rather, the declining frequency of cointegration of the selection marker and the isolated loxP site in the homologous recombination step becomes the limiting factor when a single targeting vector is used. In the TNF locus, for example, the rate of cointegration of a selection marker gene *and* a separated loxP site was found to be 70% over a distance of 4 kb, 35% over 7 kb, and only 2% over a 10 kb distance (M. Alimzhanov, unpublished results). Of course, the amount of homology remaining on the 'opposite' side of the loxP site, with respect to the selection marker, must also be taken into account. Thus, to generate deletions larger than 10 kb it is more convenient to introduce the two loxP sites in consecutive targeting events using two targeting vectors, each containing a positive selection marker gene and one loxP site (e.g. loxP-neo and loxP-hygro genes), such that these loxP sites have the same orientation when integrated into the chromosome. Alternatively to using two different selection markers, the same loxP-flanked marker gene could be used in both targeting vectors if the floxed marker is deleted after the first targeting step leaving one loxP site behind. When loxP sites are separated by a great distance, chromosomal inversions or duplications may also be generated after Cre expression in ES cells depending on the orientation of the loxP sequences (95). Attempting to delete large sequences where the gene order and orientations are not known requires a strategy such that molecular analysis of clones which have undergone Cre-mediated recombination will reveal the relative orientations of the loxP sites. Furthermore, to avoid all but the desired chromosomal rearrangements it should be confirmed that both targeting events occurred on the same chromosome. Since the efficiency of Cre-mediated deletion likely declines with increasing distance between loxP sites, a tk gene as a negative selection marker should be included within at least one of the targeting vectors (Figure 5; also see (94)). Alternatively, the reconstitution of a functional positive selection marker may be employed for the same purpose (e.g. a *hprt* mini-gene cassette upon Cre-mediated deletion in *hprt*-deficient ES cells; ref. (95)). In the mentioned experiments (94, 95) the frequency of Cre-mediated deletion upon transient transfection of ES cells was found to be in the range of 10^{-5} - 10^{-7}. Alternatively, ES cell clones containing the loxP-flanked chromosome segment could be used for the derivation of mice for mating to one of the '*deleter*' strains (96, 97) which express Cre in the early embryo, to generate the mutant harboring the deleted allele. In this case a tk gene should not be used for positive enrichment as its presence may interfere with germline transmission. However, the *deleter* strains, although performing well for the removal of small (< 5 kb) DNA fragments, have not yet been characterized for their efficiencies of deleting large DNA segments.

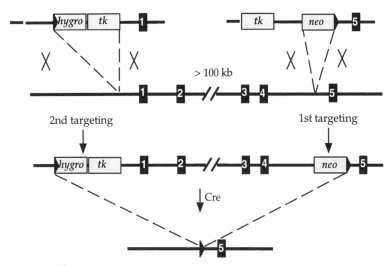

Figure 5. Generation of a large genomic deletion in ES cells by two sequential steps of gene targeting. In the first step a neo gene with one loxP site, which defines one of the endpoints of the desired deletion, is introduced into the target locus by homologous recombination. A second loxP site, defining the other end of the desired deletion, a hygromycin resistance gene and a tk gene are introduced by a second round of gene targeting into the same chromosome. Transient transfection of targeted ES cells with Cre recombinase leads to the excision of the loxP flanked DNA segment (> 100 kb) including the tk gene. Tk-negative, recombinant ES cell clones can be enriched by gancyclovir selection. The symbols are as indicated in Figure 4.

Chapter 6
Gene replacement

The design of targeting vectors for gene replacement combines features of conventional replacement vectors with the strategy to generate large deletions (see above and Figure 6). In general, the aim of such experiments is to replace endogenous wildtype sequence by a heterologous sequence (e.g. the human for murine homologue). The advantage of this strategy compared to simple insertions by conventional replacement vectors lies in the option to remove regions of the original gene, together with the selection marker used in the first targeting step, by Cre-mediated deletion. In this manner, a gene replacement vector encloses a long region of nonhomology (e.g. consisting of the intron/exon or promoter region of the gene to be introduced) and a selection marker cassette with one loxP site (e.g. loxP-neo or loxP-neo-tk). A second loxP site is placed into the homology region of the vector such that Cre-mediated deletion in homologous recombinant ES cell clones generates the desired deletion of the endogenous sequence together with the selection marker cassette (Figure 6).

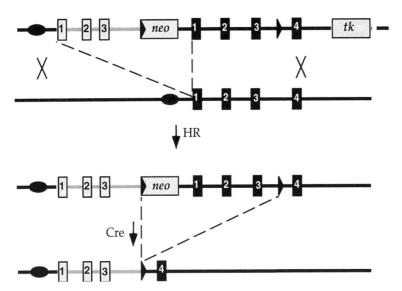

Figure 6. Cre/loxP facilitated gene replacement. In the example shown three exons, but not the promoter region, of a murine target gene are replaced in ES cells by a homologous gene segment derived from a different species (grey elements). LoxP sites are placed at one end of the neo gene and between exons 3 and 4 of the target gene. The loxP flanked gene segment and the selection marker are deleted from the homologous recombined target locus by transient Cre expression in targeted ES cells.

The size of fragments which can be inserted and deleted with this strategy is limited to lengths of up to approximately 10 kb due to the declining frequency of cointegration of the isolated loxP site with the selection marker and the impractical handling of larger vectors. If regions larger than 10 kb are to be replaced, then two consecutive steps of gene targeting are necessary as described in the section on large deletions (Chapter 5). For small replacements (< 1 kb, e.g. a single exon) it seems most convenient to directly replace the fragment to be deleted with a loxP-flanked selection marker gene attached to the inserted region (see above, Figure 4b and ref (98)). Cre/loxP-mediated gene replacement was first described for the constant region of the murine Cγ1 gene in the Ig heavy chain locus (99). In this experiment, a 3 kb fragment with six of the eight exons coding for the mouse Cγ1 gene was replaced by the homologous human DNA generating a mouse strain producing mouse-human chimeric antibodies. Cointegration of the isolated loxP site and the selection marker over a 3 kb distance occurred at a frequency of 40% in that instance. Finally, gene replacement may also be accomplished by other strategies, and in particular, in a conditional manner such that a region of DNA (e.g. wildtype sequence) is replaced in vivo with a mutated or heterologous sequence (see Chapters 9 and 11).

Chapter 7
Chromosomal translocations

To generate chromosomal translocations in ES cells using the Cre/loxP system two consecutive homologous recombination events are necessary. In the first step, a positive selection marker gene with one loxP site (e.g. loxP-neo) must be introduced by homologous recombination on one chromosome and a second selection marker with another loxP site (e.g. loxP-hygro) on the second of the two chromosomes. Alternatively to using two different selection markers, the same loxP-flanked marker gene could be used in both targeting vectors if the floxed marker is deleted after the first targeting step leaving one loxP site behind. Depending on the aim of the individual experiment the loxP sites must be placed into introns if a fusion protein should be generated or outside of transcription units to exchange only regulatory regions, e.g. mimicking naturally occurring translocations. In any case, both loxP sites should have the same orientation, with respect to the centromeres, to retain chromosomal integrity. After identification of homologous recombinants the Cre-mediated chromosomal translocation should be obtained by transient expression of Cre recombinase in ES cells (see Chapter 21). Despite the proven efficiency of Cre-mediated loxP-specific intrachromosomal deletions, a low recombination frequency for translocation events has been reported (in the order of 10^{-7} - 10^{-5}; (95, 100, 101)). Consequently, the design of targeting vectors which generate a third, new positive selection marker upon interchromosomal recombination in ES cells is generally required. To restore a selection gene by recombination, promoter and coding region of a marker gene (e.g. the puromycin resistance gene or HPRT in *hprt*-deficient ES cells) could be introduced separately in the first and second targeting steps in such a way that after Cre-mediated recombination all three marker genes will not reside on the desired translocated chromosome but on its recombination partner (and would segregate by breeding) (Figure 7). Chromosomal inversions could be generated with the same strategy as for translocations with the difference that both targeting experiments would hit the same chromosome and the two introduced loxP sites must be placed in opposing orientation. Cre has also been successfully used for recombination between chromosomes in tobacco cells (68, 102).

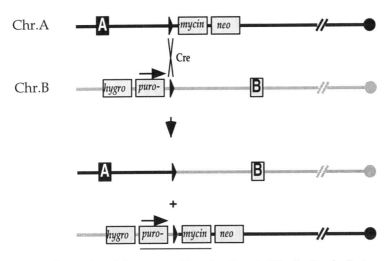

Figure 7. Generation of chromosomal translocations in ES cells. One loxP site, an intact selection marker (neo or hygro gene) and the promoter (puro-) and coding region (-mycin) of a third selection marker (puromycin resistance gene) have been introduced by homologous recombination on chromosomes A and B, respectively. The loxP sites on both chromosomes have the same orientation with respect to the centromere (filled or shaded circle). Transiently expressed Cre recombinase will exchange the arms of chromosomes A and B at the loxP sites generating a chromosomal translocation. The genetic markers A and B are brought together in the desired recombinant chromosome while the second recombination product contains the selection marker genes. The previously separated promoter and coding regions of the puromycin resistance gene are combined (thin line represents transcript) as a result of Cre-mediated recombination allowing the positive selection of recombinant ES cell clones using puromycin. The symbols are as indicated in Figure 4.

Chapter 8
Targeted integrations

Apart from generating deletions, Cre recombinase also mediates the reverse, integrative reaction by recombining loxP-containing, incoming vectors with a transgene or endogenous sequences harboring loxP sites in the genome. Thus, the capacity of the Cre/loxP system for site specific integration could be used as an alternative for homologous recombination to generate secondary modifications in loxP containing loci in cultured cells. Possible applications of loxP-mediated integration could include, for example, the analysis of different promoter regions into the same genomic site for gene expression studies or the replacement of different mutated exons in structure/function studies. For targeted integration, the loxP containing integration vector must be introduced together with a vector for transient Cre expression, ensuring only limited recombinase activity, to avoid the subsequent excision of the integrated, loxP-flanked DNA segment. In plant cells it has been shown that the equilibrium of Cre-mediated inversion can be shifted to one side by using a pair of two different mutant lox sites (42). Certainly this technique will be also used in future to stabilize the products of Cre-mediated integration in ES cells by preventing reexcision. A circular vector containing a single loxP site will be integrated completely (Figure 8), whereas from a linear vector with two loxP sites in the same orientation only the loxP-flanked region will be inserted, probably after extrachromosomal resolution of the floxed segment into a circle with a single loxP site (Figure 8). In model experiments targeted integration has been achieved in mouse, human and hamster cells (40, 57, 103), in the former case with a frequency of site specific integration of approximately 10% compared to the rate of stable transfection (similar to the frequency of homologous recombination in ES cells). Thus, Cre-mediated targeted integration seems to occur at a much lower frequency than targeted deletion (approximating 50% of *transient* transfection rate) and it is necessary to include a positive selection marker gene into the sequence to be integrated for the isolation of recombinant clones. The need to include a selection gene is, in our view, a severe disadvantage of the integration method for gene targeting applications since this marker cannot be subsequently deleted from the genome without the use of another recombinase system (e.g. FLP/FRT). The presence of a selection marker, and its regulatory regions, may disturb the genetic locus to be analyzed (20, 21, 87). Therefore, we would favor the use of conventional homologous recombination techniques even if many replacements are to be generated at the same locus, following one of the Cre/loxP based methods described above or one of the double replacement strategies (Chapter 1).

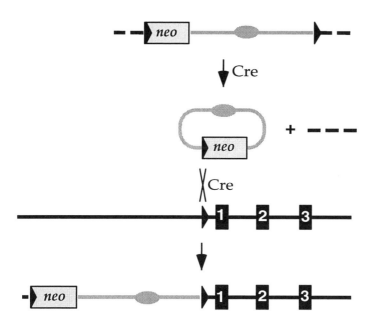

Figure 8. Cre-mediated integration. As an example the exchange of the upstream regulatory region of a gene is shown. The endogenous regulatory elements of the target gene have been deleted and replaced by a single loxP site in a previous round of gene targeting according to the 'flox-and-delete' strategy (type I deletion). Targeted ES cells are transfected with a Cre expression vector and a targeting vector containing the new, desired regulatory region, a neo gene, two loxP sites and the plasmid backbone. After Cre mediated excision of the loxP flanked vector insert the circular insert will be integrated into the loxP site of the target locus by a second recombination event placing the target gene under the control of a new promoter region.

Chapter 9
Conditional gene modification

Following one of the three strategies described in this section, from a single targeting construct, ES clones can be generated which harbor either a loxP-flanked (floxed) allele suitable for conditional gene targeting or which contain the final genetic modification resulting from Cre-mediated deletion or inversion. For the floxed allele subsequent Cre-mediated recombination will result in gene inactivation or modification, depending on loxP positioning and orientation. The mice derived from ES cells with a floxed allele can be crossed to Cre transgenic mice for conditional gene modification or inactivation in specific cell types (Chapter 11). Mice harboring the product of Cre-mediated deletion or inversion of the floxed allele in all cells throughout development, like conventional mutants, can be generated simultaneously from the second type of ES clones. Alternatively, such mice can be obtained by crossing of mice harboring the floxed allele to one of the "deleter" strains (96, 97, see Chapter 11), which express Cre ubiquitously in the early embryo.

The flox-and-delete strategy can be used for conditional gene inactivation and is especially useful if lethality is anticipated in conventional mutants. The flox-and-invert and flox-and-replace strategies do not aim to inactivate a target gene completely but are designed for the conditional replacement of a segment of a target gene segment against a modified segment. Regardless of the strategy applied, you should confirm at least the integrity of coding sequences on the final vector used for conditional gene targeting to avoid the transfer of undesired mutations into the genomic locus by homologous recombination, since a loxP-modified allele should function normally before Cre-mediated recombination. The integrity and orientation of loxP sites in vectors for conditional gene targeting should be either confirmed by DNA sequencing or in a functional assay by transformation of the construct into Cre expressing bacteria (81).

9.1 Flox-and-delete

A minimal 'flox-and-delete' gene targeting vector contains three loxP sites in the same orientation, two sites flanking a selection marker gene and an isolated site (at a distance of up to 10 kb) within one arm of homology. Thus, the construct is configured like a replacement type vector containing an isolated loxP site analogous to a nonselectable mutation as shown in Figure 9. As a floxed selection marker for vector construction we routinely use the neo gene from either plasmid pMMneoflox8 or pL2neo (Appendix B). Inclusion of the tk gene is optional, as mentioned in the previous section, given that Cre-mediated deletion occurs at high efficiency in ES cells. The single loxP site can be inserted in a restriction site in one of the homology arms of the vector either as an oligonucleotide or can be recovered from the vector pGemloxP or pGEM-30 (Appendix B).

After identification of homologous recombinant ES clones which have cointegrated the isolated loxP site (see previous section), transient Cre expression leads to three different recombination products. If recombination occurs between the outer loxP sites (type I deletion) a large deletion is created which should inactivate or modify the gene, while recombination between the loxP sites flanking the neo gene (type II deletion) leaves one loxP site in the genome which generates, together with the isolated (outer) loxP site, the floxed version of the gene (Figure 9). The third recombination product, the deletion between the isolated loxP site and the inner loxP site flanking the neo gene, is usually not of practical use and will not be found if ES colonies are screened for G418 sensitivity after Cre expression since such colonies retain the neo gene. In the experience of the Rajewsky lab ES clones harboring type I and type II deletions were found in most cases (~10) at approximately equal frequencies among the G418 sensitive colonies. In two cases, however, type I deletions were found roughly 10 and 100 times more frequently than type II deletions suggesting a differential accessibility of loxP sites for Cre recombinase at some genomic locations. If the experimental aim is to generate deletions of different size in a gene or genomic locus (e.g. deleting different exons or neighboring genes) more than one isolated loxP site can be included in the targeting vector giving rise to multiple deletion products after Cre expression in targeted ES cells. Again, when designing 'flox-and-delete' vectors, time should be devoted to identifying strategies (e.g. Southern blot digests and probes) which will be able to unambiguously distinguish between the various alleles of the targeted locus. This would be particularly important with the potential cotransfer of multiple loxP sites. A protocol for transient transfection of ES cells with Cre expression vector and for the isolation of ES clones harboring type I and type II deletions is described in Chapter 21.

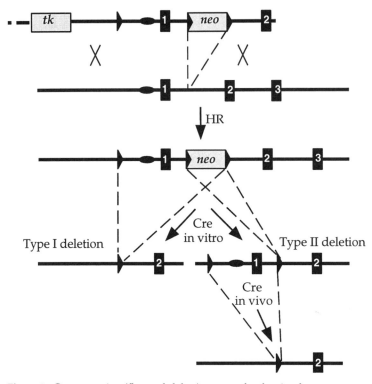

Figure 9. Gene targeting (flox-and-delete) strategy for the simultaneous generation of a loxP containing (active) and deleted (inactive) allele of a target gene in ES cells. In this example a loxP-flanked neo gene is placed in the intron between exons 1 and 2 for vector construction and a single loxP site is inserted upstream of the promoter region of the target gene. All loxP sites have the same orientation. Transient expression of Cre in targeted ES cells either leads to recombination between the outer loxP sites and the excision of a DNA segment essential for gene function (type I deletion) or to the excision of the loxP-flanked neo gene (type II deletion) generating an allele with loxP-flanked promoter region and first exon. The latter allele can be used for conditional gene inactivation by Cre expression in vivo. The third type of recombination which excises exon 1 of the target gene but not the neo gene is not shown.

ES cell clones with a type I deletion are used to produce mice harboring the inactivated version of the gene of interest and clones with type II deletion give rise to a strain with the floxed allele for conditional gene inactivation in vivo. ES cell clones with either a type I (allele) or II (neo gene) deletion also provide the option to target the second (wildtype) locus using the same targeting vector as in the first step (or a modified version if needed) as the selection marker is no longer present. The second targeting event will generate hemi- or homozygous mutant ES cells, following another round of Cre-mediated selection marker deletion, which are amenable to analysis in chimeric animals (see Chapter 21). In any case, the position of the floxed selection marker and the single loxP site within the modified gene must be carefully selected such that the gene is modified (inactivated, inverted, or replaced - see below) by recombination between the outer loxP sites, but is not disturbed by the presence of the two loxP sites remaining after deletion of the selection marker. For gene inactivation, this aim could be achieved by either flanking one or several exons within a gene by loxP sites such that a frameshift would be obtained after deletion of the floxed segment or by inclusion of the promoter region and first exon within two loxP sites. Generally it will be easier to inactivate gene function by inclusion, and subsequent deletion, of a large segment of genomic DNA within two loxP sites. On the other hand, the efficiency at which a floxed gene segment can be deleted conditionally in vivo may decrease with the increasing distance between two loxP sites although such a correlation has not been thoroughly investigated. However, efficient deletion ($\geq$ 99%) of floxed gene segments with sizes of 1.5 and 3.5 kb has been achieved in vivo (see Chapter 11). Finally, it should be noted that loxP sites have been inserted within an intron in both orientations, with respect to transcription, and homozygous floxed mice have not exhibited any overt detrimental affects with respect to RNA splicing.

The flox-and-delete strategy was first applied to generate a nonfunctional and a floxed allele of the DNA polymerase β gene in which case the promoter and first exon were flanked by loxP sites (61). A similar strategy was used to generate loxP-flanked alleles of the Interleukin-2 receptor-γ-chain (104), the N-acetylgalactosaminyltransferase (105) and NMDAR 1 (106) genes. Although only a few cases of conditional gene inactivation are published so far we are aware that a number of experiments of this type are presently underway. At our Institution the flox-and-delete approach is now routinely used to generate loxP-flanked alleles for conditional gene inactivation and it has been successfully applied to the csk, gp130, Rfx5 and immunoglobulin variable region genes (U. Betz, B. Clausen, I. Förster, K. P. Lam, W. Müller, K. Rajewsky, C. Schmedt, A. Tarakhovsky, personal communication)

9.2 Flox-and-replace

Gene inactivation is the end result of the flox-and-delete strategy, however, by a similar approach referred to as flox-and-replace a gene or gene segment can also be replaced by a modified segment in vivo to allow the synthesis of a modified protein in specific cell types (M. Kraus, R. Torres, unpublished results). Flox-and-replace allows, for example, to test for the function of single amino acids by conditional replacement but also to exchange entire protein domains.

A flox-and-replace vector (Figure 10) is similar to flox-and-delete constructs in that it harbors three loxP sites in the same orientation two of which flank a selectable marker and a third isolated loxP site, placed distal to the region of DNA to be replaced. Juxtaposed to the floxed selection marker (on the opposite side relative to the region to be replaced) is the DNA segment which will serve to replace the wildtype DNA of interest upon Cre-mediated deletion (replacement of exon 3 by a modified exon in Figure 10). The region which can be replaced using a single gene targeting vector is limited to about 10 kb due to the reduced cotransfer of the isolated loxP site (Chapter 4) and should probably not exceed 3 - 5 kb as the efficiency of Cre-mediated deletion in vivo is likely to be reduced for larger fragments. Thus, in case of genes larger than 5 kb only the 3' and 5' region will be accessible for modification by the flox-and-replace strategy. Any region of a gene can be conditionally replaced by applying the flox-and-invert strategy described in the next section.

After Cre-mediated recombination in homologous recombinant ES cells both ES clones with type I and type II deletions can be obtained by selection for cells which lost the selection marker (see Chapter 21 for protocol). Type I deletion (between the outer loxP sites) yields ES cells which replaced the DNA sequence of interest with the modified sequence. In Figure 10 this is represented by an altered exon 3 (with 3' untranslated region) replacing the wildtype exon 3 and 3' untranslated region; a similar strategy can be also envisioned for the replacement of a promoter or regulatory element 5'of the coding region controlling the gene of interest. Germline transmission from ES clones with a type I allele results in conventional mutants and provides the phenotype of the alteration in all cells throughout ontogeny. Alternatively such mutants can be obtained by crossing mice derived from type II ES clones with one of the 'deleter' strains ((96, 97), see Chapter 11). ES cells undergoing a type II deletion will harbor a wildtype gene with the exception of a loxP site within an intron and a second loxP site outside the wildtype gene followed (or preceded) by the altered DNA segment.

These alleles when transmitted through the germline, allow the analysis of the alteration in a conditional manner in vivo (see Chapter 11). An important consideration for this type of experiment is the prevention of any readthrough transcription which may complicate future analysis as a result of alternative splicing. This may be circumvented by including transcriptional stop sites (107) either upstream (for the strategy employed in Figure 10) or downstream of the loxP site placed outside the wildtype locus.

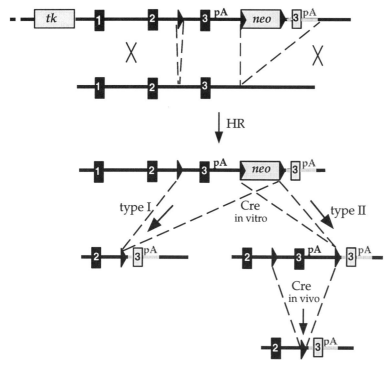

Figure 10. Flox-and-replace strategy of gene targeting results in the replacement of endogenous sequence by a heterologous DNA sequence. In this example the targeting vector has been designed to insert a floxed neo gene, together with an altered (e.g. mutated) exon 3 and 3' untranslated region (grey elements), downstream of the entire gene. A proportion of targeted ES clones will have also incorporated the upstream loxP site within the exon 2 - exon 3 intron. Transient expression of Cre, and selection for G418 sensitivity, yields homologous recombinants which have undergone either a type I or type II deletion.

9.3 Flox-and-invert

The flox-and-invert strategy utilizes the capacity of Cre recombinase to invert a loxP-flanked DNA segment if two loxP sites are placed in opposite orientation. The ultimate goal of this strategy is the replacement of a gene segment by a heterologous sequence and, therefore, is a variation on the flox-and-replace scheme. Indeed, a flox-and-invert vector is similar to the flox-and-replace vector but differs with respect to the orientation of the loxP sites to each other and the DNA region to be used in replacement to the wildtype locus. However, unlike flox-and-replace, where gene replacement is achieved by the deletion of the DNA sequence between loxP sites, with flox-and-invert gene replacement is mediated by the inversion of the floxed DNA segment (Figure 11a, 11b). The major advantages of this strategy (versus flox-and-replace) is that the configuration of the targeted locus precludes any concern of readthrough transcriptional activity and that it also can be applied to internal regions of large target genes.

A vector for the exchange of an internal exon of a target gene by Cre-mediated inversion (Figure 11a) combines the new, incoming exon with a floxed selection marker into a single heterologous region which is inserted into the intron up- or downstream of the exon to be replaced. The new exon must be part of a DNA segment which includes splice donor and acceptor sequences. Initially this new exon region is introduced in an inverted orientation in relation to the target gene such that it is spliced out of the primary transcript as part of an intron. An isolated, third loxP site must be placed into the intron region on the other side of the exon to be replaced (exon 2 in Figure 11a) in opposite orientation to the loxP sites flanking the selection marker. After homologous recombinant ES clones are identified transient transfection with Cre can result in three types of ES clones which have lost the selection marker (see Chapter 21 for protocol). Type I deletion (between the outer loxP sites, not shown in Figure 11a) yields ES cells which have lost the DNA sequences of interest and have no further value for conditional gene targeting. The two other, types II and III recombination products lose only the floxed selection marker and either contain the remaining heterologous region in the original (type II clones) or an inverted (type III clones) orientation. Germline transmission from ES clones with a type III allele results in mutants which exhibit, like conventional mutants, the phenotype of the exon exchange in all cells throughout ontogeny. By crossing of such mice to Cre transgenic strains the mutant gene can be conditionally reverted to wildtype in specific cell types. Mice generated from type II ES clones should initially express the wildtype gene product at normal levels and are crossed to a Cre expressing strain for conditional gene targeting. In double transgenic mice exon 2 of the target gene (Figure 11a) will be replaced in Cre expressing cells by the modified exon region through inversion of the loxP-flanked gene segment resulting in the inclusion of the modified exon into the mature transcript of the target gene.

As shown in Figure 11b, a similar strategy can be also followed for the replacement of a promoter or regulatory elements 5' of the coding region controlling the gene of interest, or for the combined replacement of a promoter together with one or more 5' exons of the target gene (K. P. Lam, K. Rajewsky, unpublished results).

Before planning a flox-and-invert gene targeting experiment several problems should be considered:

1. The flox-and-invert strategy may not be applied if a gene is coded on the opposite strand of the target gene since one of the inverted exons would be incorporated into this unrelated gene.

2. Since the Cre-mediated inversion reaction functions equally well in both directions an equal ratio of inverted and noninverted alleles will be finally obtained in vivo, if wildtype loxP sites are used for vector construction. The recently described strategy of using two different mutant loxP sites (42) should be useful to direct the reaction more to the product side. Depending on the configuration of the second allele of the target gene Cre expressing cells will produce only wildtype or mutant protein or a mixture of both. If the second allele is a null mutant cells will express either wildtype or mutant protein. If it is in wildtype configuration either wildtype or a mixture of wildtype and mutant protein is produced. Finally, if both alleles are modified according to the flox-and-invert strategy cells express either only wildtype (no inversion), a mixture of wildtype and mutant (one allele inverted), or only mutant protein (both alleles inverted). Such a chimeric situation may be tolerable for some but certainly not all experimental aims. In some cases, however the coexistence of mutant and wildtype cells could even be advantageous as the latter cells could serve as internal (wildtype) control. To analyze the phenotype of mutant cells in such chimeric mutants it is important to be able to distinguish cells expressing wildtype and mutant proteins, e.g. by antibodies against the wildtype and mutant proteins.

3. For conditional, but stable, gene modification mice carrying a flox-and-invert construct should be crossed only with strains which harbor a transiently inducible Cre transgene since the Cre-mediated inversion reaction is reversible. If strains would be used which express Cre constitutively inversion would continue such that Cre expressing cells would not acquire a stable, mutant or wildtype genotype.

39

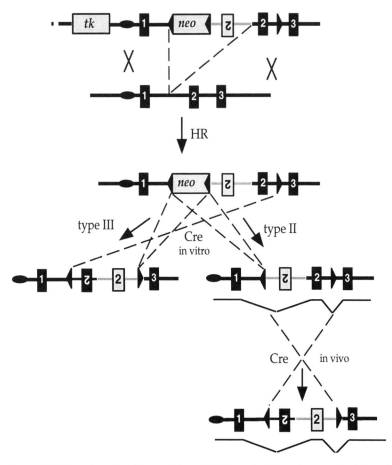

Figure 11a. Gene targeting (flox-and-invert) strategy for the exchange of an internal exon (exon 2) of a target gene by conditional gene targeting. The type III recombination is generated by simultaneous deletion of the neo gene and inversion of the loxP-flanked exon region. Thin, continuos lines beneath the genomic locus represent transcripts of the target gene. See text for further explanations.

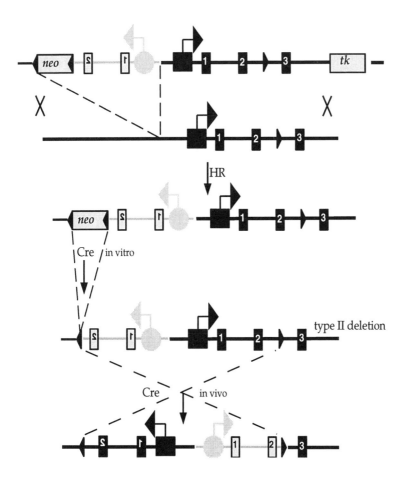

Figure 11b. Gene targeting (flox-and-invert) strategy for the exchange of a promoter region and the first two exons of a target gene by conditional gene targeting. The floxed selection marker is removed in ES cells by a type II deletion. Upon Cre-mediated inversion the promoter and first two exons can be exchanged in vivo.

Chapter 10
LoxP-containing transgenes

As discussed in the previous chapters the Cre/loxP system offers a variety of possibilities in the manipulation of targeted endogenous genes - both in the ES cell genome and mice derived from them. Moreover, Cre recombinase can be also used to activate (or inactivate) in vivo loxP-interrupted, randomly integrated transgenes introduced into the mouse germline either by pronucleus injection or the ES cell route. Similar to the conditional gene targeting strategy (Figure 13), transgene expression can be controlled by the pattern of Cre expression in vivo, i.e., in a tissue-specific or temporal manner via Cre-mediated recombination (59, 60). Especially attractive with this type of transgene manipulation is the Cre-mediated activation of (trans)gene expression. In this strategy, two transgenic strains, one expressing Cre and the other containing the recombination target must be generated and crossed together thereby allowing the synthesis of the protein encoded by the transgene only in double transgenic animals. Compared to the conventional way of controlling transgene expression directly from a cell type-specific promoter, this appears advantageous only if transgene expression is lethal or otherwise deleterious to the animal since the strain harboring the loxP interrupted (inactive) transgene could be maintained normally. Possible applications of this method include the expression of oncogenes to study tumorigenesis, toxin expression for the ablation of cell lineages and the activation of reporter genes to mark the progeny of individual cells or cell lineages. Alternatively to the production of Cre transgenic strains, a Cre expression vector can be injected directly into fertilized eggs derived from the strain with the loxP containing transgene to generate animals harboring the Cre-mediated deletion in all cells due to transient Cre expression in fertilized eggs (62). This method might be also used to generate animals with a deletion in an endogenous gene from a strain containing a gene targeting derived loxP-flanked allele although for this purpose, it is probably more convenient to cross the floxed strain to one of the *deleter* mouse strains (97, 96; see next chapter).

Transgene constructs for Cre-mediated activation of gene expression must be designed such that transcription or the production of functional protein from the expression vector is initially prevented by the inclusion of a loxP-flanked DNA segment into the vector. This dormant gene would become functional after Cre-mediated excision of the floxed spacer region leaving a single loxP site (Figure 12). Thus, it would be necessary to place polyadenylation signal sequences and/or 'transcriptional stop sites' (107) into the floxed spacer region to interfere with transcription of the transgene and/or to include an additional splice site, or a false ATG codon, to disrupt splicing or translation. Translation might be most safely prevented by separating the start codon from the rest of the coding region of the gene. This strategy would be applicable, however, only with proteins tolerating N-terminal fusions such as β-galactosidase since one loxP site would remain in the coding region of the target after recombination (see plasmid PSVlacZT, Appendix B). It should perhaps be noted here that there is only one open reading frame through the 34 bp loxP sequence.

Cre-mediated transgene manipulation has been used to activate the expression of a loxP interrupted oncogene in the eye lens by placing Cre under the control of the αA-crystallin promoter (59) and to inactivate a β-galactosidase gene specifically in T lymphocytes via Cre expression from the thymus-specific proximal *lck* promoter (60).

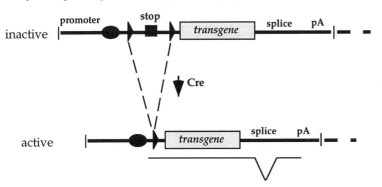

Figure 12. Cre-mediated activation of transgene expression. Promoter (filled ellipse) and coding region (shaded rectangle) of the transgene construct are separated by a loxP-flanked intervening sequence which prevents the correct transcription or translation of the transgene (filled square). This 'stop' region may contain splice donor sites, polyadenylation signal sequences and translation initiation codons as described (59). The correct transcription and splicing (thin line) of the expression construct is activated in vivo by recombinase-mediated excision of the stop sequence in cells expressing Cre. The excised reaction product is not shown.

Chapter 11
Conditional gene targeting

We understand the term 'conditional gene targeting' as a restricted gene modification in the mouse, limited to either certain cell types (tissue-specific), to only a portion of its lifetime (temporally-specific), or both (11, 12, 61). This contrasts with the conventional gene targeting strategy which creates a permanently modified allele in all cells of an animal from the onset of development (Figure 2). While the strategies discussed in this section are based on the Cre/loxP bacteriophage recombinase system, their applicability clearly extends to the other members of the λ integrase family which have the common feature of an independence on host cofactors (e.g. FLP/FRT). It must be emphasized, however, that the Cre/loxP system has already proven its usefulness and efficiency in conditional gene targeting whereas the other recombinase systems have thus far appeared less than optimal.

A requirement for Cre/loxP-based conditional gene targeting is the generation of a mouse strain which harbors a floxed DNA region of interest. This is generally accomplished by gene targeting in ES cells (e.g. following the 'flox-and-delete' strategy outlined previously) if an endogenous gene should be conditionally inactivated or modified. Additionally, expression of the loxP-modified allele should not be disturbed by the presence of the loxP sites and, upon Cre-mediated recombination, the allele should be inactivated or modified. For example, Cre-mediated loxP-specific recombination might result in gene inactivation by the deletion of a functionally essential region (flox-and-delete), or a more subtle modification may be generated by the replacement of wildtype sequence with a mutated one (flox-and-replace or flox-and-invert). This type of loxP-mediated, gene function modification would also include the conditional expression of a gene product by Cre-mediated removal of an inhibitory sequence (e.g. a floxed transcriptional termination site located between a promoter and coding sequence - see Chapter 10).

Gene modification becomes restricted in vivo in a spatial and temporal manner by crossing the floxed mouse strain to a strain expressing Cre under the control of a constitutive or inducible cell type-specific promoter (Figure 13). Obviously, these double transgenic strains must be bred homozygous, or hemizygous (heterozygous with one floxed and one deleted allele), for the chromosome containing the floxed locus.

Alternatively to the use of Cre transgenic mice the Cre gene has recently been delivered to somatic tissues of mice using Cre expressing adenoviral vectors (108, 109).

Conditional gene targeting might be particularly useful to further investigate the function of a gene product if the conventional (complete) inactivation of a gene leads to a lethal or other deleterious phenotype preventing a more detailed in vivo analysis. Furthermore, conditional targeting of widely expressed molecules can be used to test their function for a particular cell lineage. For example, a cell surface receptor present on all cells of the immune system could be selectively removed only in B lymphocytes, T lymphocytes, macrophages or other cells maintaining the majority of the immune system fully functional. The use of an inducible promoter for Cre expression has the important additional advantage that the wildtype gene product under investigation can be present throughout ontogeny until the time of induction. Inducible gene targeting should be especially helpful to analyze gene function in adults since it allows gene modification after the normal establishment of adaptational responses of cellular systems like immunological or spatial memory. This may also prevent the possible adaptation of cellular responses which may compensate if a gene product is inactivated or modified by constitutive cell type-specific Cre expression or conventional gene targeting early on during ontogeny or development. We believe this inducible approach to gene targeting will lead to a reduced frequency of 'redundant' phenotypes reported as the biological systems under scrutiny will not have had an opportunity to adapt to the genetic alteration which is the case with constitutive cell type-specific Cre expression or conventional gene targeting. The reduction of the wildtype gene product upon Cre-mediated inactivation (or modification) of its gene will of course depend on several factors which include the half-life of its mRNA and protein in the investigated cell type.

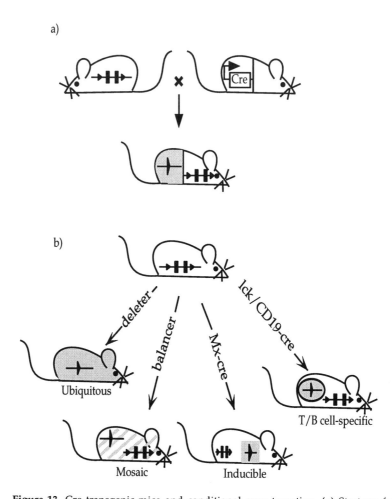

Figure 13. Cre transgenic mice and conditional gene targeting. **(a)** Strategy for conditional gene inactivation in vivo. Right mouse: a transgenic strain expressing Cre recombinase constitutively or in an inducible fashion in specific cell types or organs. Left mouse: a strain harboring a functional, loxP containing allele of a target gene generated in ES cells, e.g. as type II deletion of the flox-and-delete targeting strategy (see Chapter 9). In double transgenic animals, obtained by crossing, Cre-mediated gene modification occurs by loxP-specific recombination and is restricted to the expression pattern of the Cre recombinase (shaded area). Only one allele of the target gene is shown. The animals must be homozygous for the floxed allele to ensure the complete modification of the target gene in Cre expressing cells. **(b)** Various conditional mutants are obtained by crossing strains harboring a floxed target gene (upper mouse) to mice expressing Cre constitutively in a cell type-specific (lck-cre, CD19-cre), mosaic (balancer) or inducible manner (Mx-cre). Conventional (complete) mutants carrying the Cre-mediated deletion in all tissues can be produced using a 'deleter' strain. See also Table 1 for further Cre transgenic strains.

11.1 LoxP-flanked alleles

Cell type-specific Cre expression was first applied to classical transgene manipulation: Cre expression from the αA-crystallin promoter was used to activate a loxP interrupted oncogene construct in the eye lens (59) and to inactivate a β-galactosidase gene specifically in T lymphocytes using a Cre transgene driven by the thymocyte-specific *lck* proximal promoter (cre^{lck}) (60) (see also Chapter 10). In the latter case, the efficiency of a 3.5 kb segment deletion was determined to reach levels of up to 99% specifically in thymocytes. Conditional inactivation of an endogenous gene was first described by Gu et al. (61) for the DNA polymerase β gene in which case a 1.5 kb region containing the promoter and first exon of the polβ gene was flanked by loxP sites (polβ^{flox}) using homologous recombination in ES cells following the flox-and-delete strategy described in Chapter 9 (Figure 9). Complete inactivation of the β-polymerase gene is lethal. In polβ^{flox}; cre^{lck} double transgenic mice 63 - 84% of mature T cells, but no other cell types, were found to carry the deleted β-polymerase allele (polβ$^{\Delta}$) when the animals had one polβ^{flox} and one wild-type (polβ^{+}) allele, whereas only 38% of T cells were homozygous for the deleted allele in animals carrying one polβ^{flox} and one polβ$^{\Delta}$ allele (i.e., polβ$^{flox/\Delta}$). The incomplete deletion in polβ$^{flox/+}$, cre^{lck} mice was attributed to the transient activity of the *lck* proximal promoter, being active only during early T cell development, while the reduced deletion rate in polβ^{flox}/polβ$^{\Delta}$; cre^{lck} animals (compared to polβ$^{flox/+}$, cre^{lck} mice) is probably due to selection of thymocytes as a result of a β-polymerase deficiency (61).

Crossing of the polβ^{flox} mice to three independent strains harboring a Cre transgene (including a nuclear localization signal sequence) driven by a constitutive active B lymphocyte-specific promoter resulted in 100%, 20% and 5% deletion of the floxed gene segment in B cells of animals with a polβ^{flox}/+ genotype (R. Rickert and K. Rajewsky, unpublished results). The various efficiencies of in vivo deletion in these animals is most likely attributed to the different expression levels of Cre in the individual transgenic lines as a result of differences in transgene copy number and genomic integration site. Indeed, it is our experience that complete deletion of a floxed allele is certainly feasible but requires the identification of an appropriate Cre transgenic line indicating that often a number of founder lines should be examined when attempting to express Cre as a transgene. However, additional work is required to determine whether the efficiency of Cre-mediated deletion also varies between different target loci e.g. in relation to chromatin structure, transcription rate or DNA methylation. In cell lines the efficiency of Cre-mediated insertion was shown to vary greatly on the chromosomal position of the target (110).

Since we have not found any unexpected recombination products of the floxed β-polymerase locus in multiple different Cre transgenic strains we

47

assume that the mouse genome does not contain cryptic, loxP homologous sequences which could be efficiently recombined with genomic loxP sites introduced by homologous recombination. As this assumption is based on the limited sensitivity of Southern blot analysis these results do not exclude, however, recombination between a floxed allele and other loci at low frequency as found in yeast (111).

A second gene product, N-acetylgalactosaminyl-transferase, that is widely expressed and embryonic lethal when inactivated, has also been conditionally ablated in the T cell lineage (105). This polypeptide catalyzes the transfer of the monosaccharide GalNAc to serine and threonine residues during O-linked oligosaccharide biosynthesis. Thus, by crossing mice harboring a floxed exon, critical for catalytical activity, to cre^{lck} transgenic mice, T cell development and peripheral colonization was assessed in the absence of N-acetylgalactosaminyl-transferase activity (105). In addition, the flox-and-delete strategy was also used to generate loxP-flanked alleles of the genes for Interleukin-2 receptor γ-chain (104), low-density lipoprotein receptor-related protein (LRP) (108) and NMDA receptor 1 (106). Although only a limited number of conditional gene targeting experiments have been published so far we are aware that a number of experiments of this type are presently underway. To our knowledge, the flox-and-delete strategy has been successfully used for the conditional inactivation of the Csk, Gp130, Rfx5 and immunoglobulin variable region genes (U. Betz, B. Claussen, I. Förster, K. P. Lam, W. Müller, K. Rajewsky, C. Schmedt, A. Tarakhovsky, personal communication).

11.2 Cre transgenic mice

Cre transgenic mice can be produced either by pronuclear microinjection of conventional, randomly integrating transgene constructs (112, 113), by targeted transgenesis (114, 115) or by the targeted introduction of the Cre recombinase gene in frame within an endogenous coding region. The former approach is much more straightforward considering the effort involved in vector construction and its introduction into the germline (112 for the production of transgenic mice, 113). However, a promoter region tested for transgenic expression must be available and the number of different founder lines which must generated is somewhat unpredictable in order to identify a useful strain as the level and pattern of transgene expression often varies greatly depending on its copy number and integration site(s) (116). In any case, the expression pattern of the promoter region used will determine the onset and cell type specificity of Cre-mediated gene modification, while the expression level determines the efficiency of gene modification in a given cell type. Presently, the exact relationship between the number of Cre molecules in a cell and the extent of deletion of a floxed gene segment has not been established. In addition, the efficiency of Cre-mediated recombination seems to depend on the chromosomal position of the target as shown for integrative recombination in cell lines (110). For the identification of a Cre transgenic strain suitable for deletion experiments it is presently necessary to produce a series of transgenic mice with one type of construct and to compare the deletion efficiency among these lines empirically by crossing them to an indicator strain possessing a floxed gene segment. The extent of deletion in tissues or cell types can be either directly assessed at the DNA level by Southern blot detection or at the level of single cells by detecting β-galactosidase activity in histological sections if mice transgenic for a β-galactosidase recombination vector are used as the indicator strain (62). A monoclonal antibody specific for the Cre recombinase has recently been generated and will clearly be useful for screening Cre transgenic lines as it is proficient for both Western blotting and flow cytometry (117).

Transgene constructs for expression of Cre recombinase can be designed like common cDNA expression vectors for transgenic mice, i.e. a promoter region coupled to an intronless Cre coding sequence which is followed (or preceded) by a region providing splice donor/acceptor sites (to increase cDNA expression (118)), and a polyadenylation signal sequence (Figure 14a). For pronuclear microinjection the entire expression unit should be recovered from the vector as a fragment using unique restriction sites to avoid cointegration of plasmid sequences. The selection of a promoter region for Cre expression will depend on the purpose of the investigation. For vector construction, plasmid pGKcre^{NLS}bpA (Appendix B) can be used to isolate the Cre coding region including an N-terminal nuclear localization signal and Kozak consensus nucleotides for efficient translation (see ref. (58) for sequence). The presence of the nuclear localization signal has not been thoroughly examined for its increased efficiency in Cre-mediated deletion and, in theory, the 38 kDa Cre protein should pass through the nuclear membrane even without this localization signal. As elements providing splice donor and acceptor sites, we have successfully used intron/exon regions derived from the β-globin (118) or the human growth hormone gene (119). Before the actual production of transgenic animals, vectors designed for Cre expression should obviously be tested for recombinase activity in cell lines. We generally test such vectors by transient cotransfection with the recombination substrate vector pSVlacZT (Appendix B), expressing β-galactosidase after Cre-mediated deletion of a floxed spacer region, into a cell line known to activate the promoter chosen for Cre expression. The intact version of the β-galactosidase expression vector should be transfected in parallel as a positive control to confirm the activity of the promoter driving the β-galactosidase gene in the cell line used and to determine the efficiency of transient transfection. A protocol for assaying β-galactosidase activity in cultured cells is provided in Appendix C.

For constitutive, cell-type-specific Cre expression, naturally occurring or engineered promoter/enhancer combinations can be tested for suitability to conditional gene targeting. For the temporal control of Cre, its activity can be presently regulated at both the transcriptional and posttranslational

49

levels. Cell-type-specific transcriptional control can be achieved by one of the systems which regulate the activity of engineered minimal promoter regions by specific transactivating proteins which are, in turn, regulated by the presence of an inducer molecule like tetracycline (120) or ecdysone (121). By placing the Cre gene behind a minimal promoter and directing cell-type-specific expression of the transactivator, site-specific recombination can be controlled in mice by the administration of inducer (122). Both regulatory systems have been shown to efficiently control the activity of reporter genes in transgenic mice (121, 123, 124). A disadvantage of these systems is that they require independent introduction of the genes coding for the transactivator and Cre recombinase into the genome, thus large numbers of transgenic mice must be crossed and tested to identify strains in which both genes are optimally regulated. This opinion, however, has been recently challenged by the demonstration of efficient regulation of a reporter gene in transgenic mice containing the reporter and transactivator genes on a single vector (125). Posttranslational control of the activity of site-specific recombinases can be achieved in the form of a fusion protein of the recombinase and the ligand binding domain (LBD) of steroid receptors expressed by a constitutively active cell-type-specific promoter. In the absence of hormone the steroid receptor LBDs are bound by heat shock proteins which inactivate the recombinase, presumably by steric hindrance (see ref. (126) for review). The recombinase can be activated by the addition of hormone which releases the heat shock proteins from the fusion protein. In the initial experiments by Logie and Stewart (73), the activity of FLP recombinase was shown to be regulated by fusion with the LBDs of wildtype steroid receptors. To derive a system in mice which is unresponsive to natural steroids, Cre was fused to mutant LBDs of the estrogen (127, 128) and progesterone receptors (129) which are unresponsive to their natural ligands but can be activated by synthetic hormone antagonists. By using a fusion protein of Cre and a mutant estrogen receptor Cre can be induced to delete a floxed gene segment in transgenic mice upon administration of the inducer 4-hydroxytamoxifen (128). Induced deletion can be targeted to lymphocytes by expression of a similar protein in B cells (F. Schwenk, unpublished results).

a) Cre expression vector

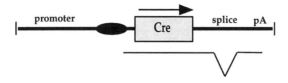

b) Targeted insertion of Cre

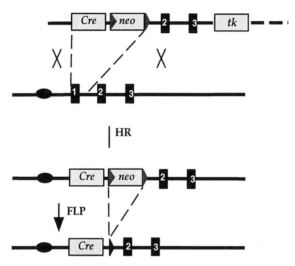

Figure 14. (a) Diagram of a Cre expression cassette as used for the production of transgenic mice by pronucleus injection. A promoter region, which has been previously characterized for gene expression in transgenic mice, directs transcription of the 1 kb coding region of Cre recombinase (shaded rectangle). Splice donor and acceptor sites are placed in the 3' untranslated region for efficient transgene expression followed by a polyadenylation signal sequence (pA). The resulting transcript is drawn as a thin line. **(b)** Diagram of a vector for the targeted insertion of Cre. The first exon of the target gene is replaced in the targeting vector by the coding region of Cre such that the position of the translation initiation codon is unchanged. A neo gene flanked by FRT (or mutant loxP) sites (stippled triangles) is inserted downstream of Cre. The FRT (mutant loxP) flanked neo gene must be deleted from the targeted locus by transient expression of FLP (Cre) recombinase. The symbols are as indicated in Figure 4.

As an alternative to the transgenic approach, the Cre gene can be brought under the control of an endogenous promoter by homologous recombination in ES cells, i.e., replacement of the endogenous gene with the Cre gene. By this strategy Cre expression becomes optimally regulated since all control elements of the targeted gene are present at their natural chromosomal position avoiding the problems of inappropriate transgene expression often encountered with conventional transgenic mice (116). Furthermore, the targeting approach can be applied to any gene for which genomic clones are available without the need for characterized promoter regions necessary for transgenic expression. Gene targeting is, however, more laborious and time consuming as compared to the construction of conventional transgenic mice. Furthermore, the generation of a nonfunctional allele of the gene used for targeted Cre expression might be another disadvantage in certain cases as it would then be necessary to maintain it in a heterozygous state and, thus, at a reduced expression level. Hence, given the availability of a well characterized promoter, we would favor the former approach for transgenic Cre expression.

Vectors for targeting Cre to an endogenous promoter region should be configured as replacement vectors to avoid the genomic integration of plasmid sequences, and their potential negative influence on gene regulation, with insertion type vectors. A vector for targeted Cre insertion must be constructed such that the initiation codon of the targeted gene is replaced by the coding region of the Cre gene to be expressed (Figure 14b). For use in vector construction, the Cre sequence can be either directly isolated from either of the plasmids pGKcre^{NLS}bpA or can be recovered by PCR using a 5'-primer including the nuclear localization signal sequence (58). We would suggest not placing an additional polyadenylation signal at the end of the Cre gene thereby allowing transcription to proceed through the intron/exon region of the targeted gene to improve Cre expression and to preserve proper gene regulation. A selection marker gene must be included downstream of the Cre sequence within the targeting vector for the identification of targeted ES clones which should later be removed from the genome to minimize the disturbance of the targeted locus. We prefer not to use a loxP-flanked selection gene for this purpose to avoid potential chromosomal rearrangements since three loxP sites would be present in the genome if a loxP containing Cre expressing mouse strain would be crossed to another strain possessing a floxed gene. Thus, for targeting vectors which insert Cre, the selection marker should be flanked either with FRT sites to remove the marker gene in recombinant ES clones using FLP (20, 69, 70) or, alternatively, the selection marker can be flanked by mutant loxP sites (41), which do not recombine with wild-type loxP sites, and transient expression in ES cells of the Cre recombinase (Chapter 21). In the first example of targeted insertion Cre was placed into the CD19 locus for expression in B lymphocytes (130, 131).

Presently, as summarized in Table 1, a limited number of Cre transgenic strains is available, allowing conditional gene targeting in specific or many cell types, either constitutively or upon induction. The initial experiment using the proximal lck promoter for Cre expression in thymocytes (lck-cre) to delete a loxP-flanked segment of the DNA polymerase β gene showed that, in principle, cell-type-specific targeting of an endogenous gene is feasible (61). Using a newly derived lck-cre strain complete deletion of a floxed allele of the GalNac-T gene in thymocytes was recently achieved (105). In the first example of inducible gene targeting Cre was placed under the control of the Interferon-α/β inducible promoter of the Mx1 gene (Mx-cre) (63). In one transgenic line generated, complete deletion of pol$\beta^{flox/flox}$ or polβ^{flox}/Δ alleles is achieved in the liver and lymphocytes two days after treatment with interferon indicating that induced Cre-mediated deletion can proceed rapidly and also efficiently in an organ composed mainly of resting cells. These mice also showed that two copies of a floxed allele in homozygous mice can be as efficiently deleted as a single, heterozygous allele. Neuron specific gene deletion can be obtained by expression of Cre from the promoter of the calmodulin-dependent kinase II (ref. (132) and R. Kühn, unpublished results). Partial gene inactivation in all tissues of the body can be achieved using the *'balancer'* strain expressing Cre under the control of regulatory elements of the nestin gene (133). This strain can be used to study competition between wildtype and mutant cells since all organs are chimeric for these two cell types. To delete a loxP-flanked gene segment from the mouse genome, e.g. for removal of a selection marker or gene inactivation, mice can be crossed to strains which express Cre during early embryogenesis (*deleter* strains) so that all offspring contain the Cre-mediated deletion in germ cells (96, 97).

Because of the potential offered by the Cre/loxP recombination system, we assume that presently many tissue-specific Cre transgenic mouse strains have been, or are being, developed. With respect to the substantial energy required to generate and characterize Cre-expressing strains, it would be useful to establish an international network for the exchange of such strains and information.

Table 1. Cre expressing mouse strains (December 1996).

Promoter/enhancer region	Cre expression	Reference
Proximal lck promoter	Thymocytes	(60, 105)
A-crystallin promoter	Eye lens	(59)
CD19 promoter	B lymphocytes	(130)
CamKII promoter	Hippocampal neurons	(132)
Nestin promoter/enhancer	Embryo, mosaic	(133)
Mx1 promoter	Ubiquitous, inducible	(63)
CMV promoter	Ubiquitous, inducible	(128)
CMV minimal promoter	Early embryo (*deleter*)	(97)
Adenovirus EIIa promoter	Early embryo (*deleter*)	(96)
HSV-tk promoter/enhancer CMV promoter	Cells infected with adenoviral vector	(108, 109)

Section 2
Working with ES cells

Chapter 12
ES cell transfection overview

Thinking about addressing a biological question by gene targeting is an exciting aspect of the experiment and one that requires relatively little knowledge about ES cells and the practicalities of gene targeting. Similarly, constructing a targeting vector can also be accomplished with little knowledge of ES cells. The time when ES cells are often first considered, at least in a practical sense, is when the targeting construct is completed and you are ready to introduce the DNA into the ES cells in attempts at homologous recombination. Without any ES cell experience it is at this point that a slight trepidation is often experienced. This feeling is enhanced for the newcomer by the apparent confusion which surrounds the juggling of *EF* cells for each *ES* cell manipulation. This section attempts to provide an overview of these manipulations, including the transfection of ES cells, followed by detailed protocols which we have used in generating homologous recombinant ES cell clones and the transfer of the altered allele to the germline. Of course, once your mutation has been transmitted to the germline, the breeding, maintenance, and analysis of the mouse strain is necessary. The topic of mouse genetics is not addressed here but we refer you to an excellent book on this topic (134).

Working with ES cells requires you to think ahead, primarily because ES cells are generally cultured on mitotically inactive embryonic feeder (EF) cells instead of directly on (gelatin-treated) plastic (there are exceptions described below). So, if you rely on mitomycin C treatment to mitotically-inactivate EF cells, then this should be done one day in advance of any ES cell manipulation. For example, this would mean that if you were planning on manipulating (e.g. passing or transfecting) ES cells on Wednesday then on Tuesday there should be enough confluent plates of EF to mitomycin C treat so that they would be ready for the ES cells the following day. Therefore, the EF cells should have been previously expanded (e.g. on the weekend) such that there would be enough confluent plates on Tuesday to treat (Figure 15).

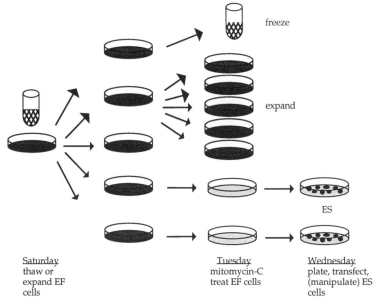

Saturday	Tuesday	Wednesday
thaw or	mitomycin-C	plate, transfect,
expand EF	treat EF cells	(manipulate) ES
cells		cells

Figure 15. Expansion and mitotic inactivation of EF cells prior to manipulation of ES cells.

We can extend this example by assuming that on Wednesday you transiently transfect a homologous recombinant clone with the Cre recombinase and plate these transfectants on the mitomycin C-treated EF cells (see Section 1 for details on Cre recombinase/loxP recombination). Accordingly, on Friday you will need an additional 2 or 3 plates of mitomycin C-treated EF cells to replate the Cre transfectants on in order to obtain well separated colonies for picking in about ten days (this will be explained later). The EF cells expanded (or thawed) on Saturday can be used efficiently for both the plating of transfected cells on Wednesday and the replating of these transfected cells which would be necessary on Friday. This would most conveniently be accomplished by splitting one or two of the original five plates on Tuesday in addition to mitomycin C-treating the two or three plates for the plating of transfectants. This would result in the expanded EF plates being confluent on Thursday which could then be mitomycin C-treated and ready on Friday for replating. Thinking about **EF** cells every time you want to do something with **ES** cells eventually becomes second nature but at first is often overlooked.

This chapter presents a very general outline of a typical targeting experiment (Figure 16) assuming you have a targeting construct in hand (or very close to in hand). The purpose of this overview is to give a general idea of the time frame and requirements, with respect to EF and

ES cells, when transfecting ES cells. More detailed protocols for each step are provided in following chapters.

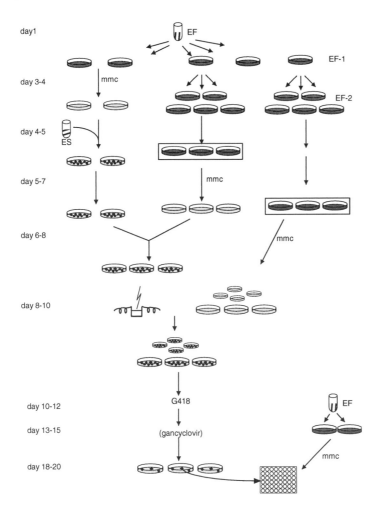

Figure 16. ES cell transfection overview. Approximate time frame:
day 1:thaw EF onto 4 or 5 plates (EF-1)
day 3-4:mmc treat 2 EF plates and expand rest (1:5) or freeze some down
day 4-5:thaw ES onto 2 mmc-treated EF-2 plates
day 5-7:mmc treat 2 or 3 EF-2 plates
day 6-8:pass ES for transfection
day 7-9:mmc treat enough EF plates for transfected ES + controls
day 8-10:transfect ES
day 10-12:G418 selection (48 hour post transfection)
day 13-15:(gancyclovir selection); thaw new EF for colony picking
day 17-19:mmc treat EF for colony picking (48 well plates)
day 18-20:pick colonies (PCR screen)

At this point, grow up ES in 48 well plates until sufficient numbers to freeze down plates

1. Thaw embryonic feeder cells (from your previously prepared stock), expand, and mitomycin C-treat prior to thawing ES cells. These feeders can be expanded according to need and can even be frozen down again after one or two passages. One week prior to picking ES colonies (i.e., 1-2 days after G418 selection) <u>may</u> need to thaw new EF cells depending on their (EF cells in culture) condition, number of transfectants, etc.

2. Thaw ES cells and plate out onto mitomycin C-treated EF cells. Depending on the number of ES cells frozen down/vial, may need to plate out onto more than one dish of treated EF cells. Assume 50% of ES cells will not survive and at maximum do not want to exceed ~1 x 10^7 ES cells/ 9 cm plate. If you do not want to have to pass these ES cells immediately, then plating on to two dishes of treated EF cells is perhaps better. But be careful as a minimal density of ES cells is required to prevent an increased frequency of differentiation.

3. Feed ES cells with fresh DMEM-ES media the following day. Expand as needed, plating ES cells at ~10^6 cells/9 cm dish of confluent mitomycin C-treated EF cells. Generally, ES cells plated at this density will require several days before passing is needed again. Two days before transfection, plate ES cells at higher density than normal at ~2 (or 3) x 10^6 cells/9 cm dish.

4. Mitomycin C-treat confluent EF cells one day before either expanding ES cells or transfection. Treat enough EF cells for plating transfected ES cells **including controls**; e.g. for 2 x 10^7 transfected cells, plate ~3 x 10^6 / 9 cm dish (x 6 dishes) and 1 x 10^6 / 6 cm dish (x 2 for control dishes); also include a plating efficiency dish.

   Controls: 1) Plating efficiency.

   2) G418 resistance (transfection efficiency).

   3) Gancyclovir enrichment (when using tk as a negative selection marker).

5. Day of transfection feed early and then wash ES cells in (MT)PBS (mouse isotonic PBS) and trypsinize. Aliquot 10^7 cells/15 ml tube, spin down and resuspend in 1 ml transfection buffer containing ~20-30 μg linearized DNA (per 10^7 cells transfected). Although PBS is adequate, we prefer to use MT-PBS for our ES cell work.

6. Aliquot 1 ml to electroporation cuvette and electroporate. After electroporation, add to ES medium (allow to sit for 5-10 minutes) and plate out at ~3 x 10^6 cells/9 cm dish (maximum 5 x 10^6 cells) plus controls.

7. Forty-eight hours after transfection exchange medium with fresh ES medium-containing G418. Feed daily thereafter with selection medium.

8. If thymidine kinase was utilized in your targeting construct to select against random integrants, then 5 days after transfection (3 days after G418 selection) start gancyclovir selection by the addition of freshly prepared gancyclovir to G418 medium. Continue double selection until colonies are picked.

9. Approximately ten days after transfection, dishes should be relatively clean (i.e., no cellular debris from dying colonies) with resistant colonies clearly visible. At this time colonies are harvested by transferring them under a dissecting scope in an outward pressure tissue culture hood.

10. Isolated colonies are expanded in 48 well plates on mitomycin C-treated EF cells. If PCR screening, half the colony is placed in PCR tube for amplification and if screening by Southern, half the colony is placed in a duplicate 48 well plate for expansion.

Chapter 13
Embryonic feeder cells

To maintain their pluripotency, ES cells are usually cultured either on plastic plates (optionally can be gelatinized) in the presence of leukemia-inhibitory factor (LIF) or on a layer of mitotically inactive feeder cells derived from embryonic fibroblasts. These embryonic feeder, or EF, cells are prepared from day 13-14 embryos. We culture our ES cells on mitomycin C-treated primary EF cells *and* in the presence of LIF.

Preparation of EF cells obviously needs to be done in advance of any ES cell work. EF cells must be G418-resistant (or resistant to any other selection drug you wish to use in isolating homologous recombinants) and, therefore, must be prepared from mice harboring the neo gene. We use a 129-derived strain which harbors pSV2neo (135).

Preparation

1. Set up breeding pairs of males homozygous for the neomycin resistance gene crossed to e.g. BL/6 females. Should plan ahead to ensure sufficient quantities and/or appropriate age of the appropriate transgenics are available.

2. Check for vaginal plug the next day and separate plugged females; this is day one. Embryonic feeder cells are prepared on day (13 or) 14.

3. On day (13 or) 14, sterilize surgical instruments (several fine forceps and scissors) and leave in hood. Prepare several dishes of sterile (MT)PBS for washing embryos (use non-coated dishes). Sacrifice pregnant mice and rinse well with 70% EtOH. Pin down and bring to hood.

4. Cut back skin, peritoneal wall and locate embryos. Cut away and place uterus into sterile (MT)PBS. Wash uterus several times by transferring to new (MT)PBS dish.

5. Using two fine forceps, disrupt uterine wall and free embryos. Remove fetal liver and heart (red parts) placing embryo in new (MT)PBS dish. Cut away the head placing the rest of the embryo in a new dish in a minimal amount of (MT)PBS. Finally, cut into small pieces.

6. Transfer embryos together with 50 ml 1x trypsin/EDTA into an Erlenmeyer flask with glass beads (4-5 mm; washed and previously heated at 170°C) and stir very gently at 37°C for 30 minutes.

7. Add EF medium to flask and transfer to 50 ml tube(s). Wash beads with medium. Spin down cells and resuspend in fresh medium and count., e.g., for 4 mice and ~15 embryos then ~2 x 10^8 cells should be resuspended in 700 ml.

8. Plate in 9 cm dishes as needed. Assuming 50% of cells will not adhere a minimum of ~1 x 10^6 cells/10 ml/dish should be plated. The upper limit will depend on how many dishes you would like to maintain and when you would like to freeze them down. Plate in 10 ml of medium which should suffice for $\leq$ 4 days or until confluent. One plate should be tested for neomycin resistance by addition of 1 mg/ml G418 the following day.

9. Trypsinize a confluent plate and freeze in 10% DMSO in DMEM-EF (see appendix for DMEM-EF) medium. One plate/vial should be ~5 x 10^6 cells.

EF cell tissue culture

You should have an approximate idea about what your EF cell needs will be when you thaw a fresh vial which means how much (and what kind) of ES cell manipulation will be done. Since a frozen vial of EF cells is generally sufficient to plate into five 9 cm dishes, after a second passage you can potentially generate 25 confluent 9 cm dishes of EF cells or >10^8 EF cells. The point is that you will be unlikely ever to have such a need for so many EF cells at one time and, therefore, you may want to freeze down different passages to be used later (see Figure 15). Alternatively, you may leave a confluent EF dish for a day or two before using thereby staggering the passing of the EF cells.

1. Thaw one tube of EF cells (~5 x 10^6 cells) into five 9 cm dishes with ~10 ml DMEM-EF medium/plate. Thawing cells from liquid N_2 should be done as quickly as possible by agitating in 37°C water bath (clean tube with alcohol before opening) and immediately adding cells to ~10 ml warm (or room temp) medium. Centrifuge at ~1200 rpm for 5-10 minutes and resuspend in medium. This removes DMSO (from freezing medium). Plate out cells.

It is imperative that tissue culture plates are used (i.e., especially for tissue culture (t.c.)) as EF cells will not adhere to non-tissue culture plates.

2. Once confluent (3-4 days), split cells. Depending on the particular requirements, you can split 1:4 or 5 for expansion, or less if confluent cells are needed sooner. Generally, expand EF cells by plating at ~10^6 cells/9 cm dish. A confluent 9 cm dish has ~5×10^6 cells although this will vary with the number of EF cell passages (a confluent earlier passage EF plate will be more dense than a confluent late passage EF plate).

It is a good idea to number passages (e.g. EF-1, EF-2) as to have an idea about the number of EF cells/confluent dish. Generally, EF cells are not passed more than 3 times (EF-3). Of course in certain circumstances EF-4 may be used, such as passing ES cells for one day until fresh EF cells are obtained. When possible, however, EF-4 are avoided.

If ES cells are to be cultured on a single plate for more than several days (e.g. during selection after transfection) then early passage EF cells are preferable. With extended time in culture EF-3 (and later passage) cells do not do as well as earlier passage EF cells and have a tendency to become detached from the plate when exchanging medium or, worse, when picking colonies in (MT)PBS.

Trypsinization

1. Dilute 10x Trypsin/EDTA to 1x with (MT)PBS and filter sterilize and store at 4°C.

2. Remove media and wash once with (MT)PBS. Add 3 ml 1x Trypsin/EDTA to a confluent 9 cm dish and place at 37°C for 2-3 minutes or until cells round up (use 1 ml trypsin /confluent 6 cm dish; 0.5 ml / 24 well plate; 0.1 ml / 48 well plate). During incubation add 7 ml of EF medium to 15 ml conical tube. Once cells have detached, pipette gently in dish to break up aggregates. If clumps are still present then place back at 37°C until fully trypsinized. Add trypsinized cells directly to medium and centrifuge at ~1200 rpm for 5-10 minutes. Count cells in hematocytometer while spinning cells down. Plate at ~10^6/dish for expansion.

Mitomycin C (mmc) treatment

An alternative method for mitotically inactivating EF cells is gamma irradiation. Although irradiation may be quicker and more convenient, the facility required for such treatment is not always readily available. Consequently, the straightforward method of incubating the cells in the presence of 10 µg/ml of mitomycin C is presented in this manual.

1. For a confluent plate of EF cells, wash with (MT)PBS and add 4 ml of mmc-containing DMEM (10 μg/ml; 1:100 of stock - see appendix for mmc preparation).

2. Incubate at least 2 (up to 6) hours in incubator at 37°C.

3. Wash 3x with (MT)PBS and then add 3 ml 1x trypsin/EDTA. Incubate at 37°C for 2-3 minutes or until cells round up.

4. Gently resuspend cells in plate until single cell suspension (if still aggregates, place back at 37°C for minimal time required to fully trypsinize cells). Transfer immediately to 7 ml DMEM-EF medium.

Chapter 14
ES cells - handling and use

Pluripotent embryonic cells (ES cells) were first established directly from murine blastocysts in 1981 (136, 137) after unsuccessful early attempts to obtain germline chimerism with teratocarcinoma (EC) cells (138, 139 for review), and later shown to efficiently produce germline chimeras (140) even after transfection (141, 142). Since then, a fairly large number of ES cell lines have been generated and the lines currently often used for gene targeting experiments include D3 (143), E14 (144, 145), AB1 (146), R1 (147), J1 (148) and CCE (140) which are all derived from 129 substrains or 129 hybrids. A germline competent *hprt*-negative line which can be used with HPRT as selection marker has also been described (149). There is a smaller number of less frequently used ES lines derived from other strains like ES623 (143) and B6-III (150) which were established from C57BL/6 mice. For the derivation and establishment of new ES cell lines see Robertson (14) and Abbondanzo, et al. (151).

The choice of which ES cell line to use may be dictated by factors such as the genetic background necessary for your experiments or the source of DNA used in the targeting construct. Ultimately, however, the goal is to deliver your genetic perturbation to the germline. The appropriate genetic background can always be obtained by backcrossing (although often at a heavy price in terms of time) and the frequency of homologous recombinants might be low with non-isogenic DNA but a higher frequency of recombinants will not be of any use if you do not get germline transmission. Hence, for a successful gene targeting experiment it is important that you use ES cells which have demonstrated the ability to colonize the germline at a high frequency. We have primarily used the E14 and B6-III lines and what follows are a few guidelines for the culture of these cells.

The ES cells which you culture in the incubator can participate in the development of a mouse. This is an incredible fact to conceptualize as you begin to work with these cells - regardless of your tissue culture experience, these cells will look different in some way. This feeling may dissipate with time and familiarity but almost always returns with your first chimera and again with germline transmission. While it is true that ES cells are somehow special cells, they are also simply tissue culture cells which need to be fed, passed, and maintained. The danger here is that mistreatment of ES cells will easily lead to differentiation from this pluripotent developmental state. Consequently they may be unable to participate in development and, therefore, chimerism and/or germline transmission will not occur. The primary goal in handling ES cells is to preserve their pluripotency with respect to embryonic development. Fortunately, this can easily be obtained by careful tissue culture technique.

Unlike some tissue cultures, ES cell maintenance cannot be postponed. ES cells should be treated as optimally as possible with constant regular feedings, i.e., not waiting for the medium to change color. If culturing high concentrations of ES cells on a single dish is required for a particular reason (e.g. prior to transfection or when waiting for mmc-treated EF), it would not be unusual to feed these cells early in the morning and again in the evening. When culturing ES cells, the cell density is very important - cultures should be split frequently (every 2-3 days) such that individual colonies do not become too large (<1000 cells). Similarly, when trypsinizing ES cells it is important to fully resuspend the cell clumps (by gentle pipetting) such that no large aggregates are transferred to the new plates. The reason for these precautions is that large colonies easily differentiate and likely lose their pluripotency. Also, the surface area of the tissue culture dish covered by colonies should not exceed ~50% since ES cultures will completely differentiate into endoderm-like cells within a few days if the cultures have ever been confluent (Figure 17).

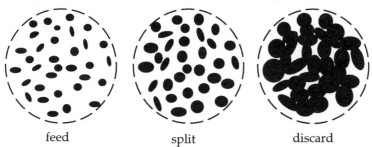

feed   split   discard

Figure 17. ES cell densities

DMEM-ES should be carefully and thoughtfully prepared making sure to include all the supplements at the correct concentrations and especially careful to use the appropriate FCS (see below). It is important to understand what good, undifferentiated ES cells look like in terms of morphology and growth characteristics. This will vary slightly with different ES cell lines, e.g. our E14 colonies generally appear more oblong than the B6III which are more round. However, it is best to learn this first hand and examples of undifferentiated vs differentiated cells should be pointed out to you. If possible it's a good idea to actually handle and familiarize yourself with these cells prior to your actual experimental need.

If ES cells are cultured appropriately the fraction of overtly differentiated cells, which can be easily recognized in the microscope by their differing (e.g. fibroblast-like) morphology and size, is usually very low. This kind of overt differentiation is not problematic since such differentiated cells have a limited lifespan and are continuously diluted out by splitting the cultures. However, even if a culture has a uniform appearance we imagine an ES cell line as a heterogeneous population of cells differing in their ability to colonize the germline in chimeras. This idea originates from the common observation that wildtype (or targeted) ES cell (sub)clones can be identical with respect to cell morphology and their capacity to generate coat color chimeras, but differ in ability to participate in germline development. Thus, a typical ES cell population (cell line) as used for transfection probably contains germline-competent cells, but also a varying fraction of cells which are, for unknown reasons, still able to form somatic, but not germ cells in chimeras. The fraction of cells which have lost pluripotency may increase with passage number and may differ from one isolate of a line to another. Such cells cannot be currently distinguished from germline-competent ES cells by morphology, karyotype, in vitro differentiation, or any other in vitro test. Therefore, prior to using for gene targeting experiments, the germline competence of an ES cell line might be confirmed by the production and test breeding of chimeras. Testing ES cell lines, however, may be misleading since it simply confirms the presence of a certain fraction of germline competent cells within the population and may nevertheless result in the frustrating experience that the majority of gene targeted clones derived from such a line are unable to contribute to germ cells. In our own experience, for example, a certain isolate of D3 ES cells exhibited a good record of germline transmission (80% of chimeric males) but, when transfected, only about 1/5 of targeted clones were able to contribute to the germline (at a high rate), although all clones (> 20 clones tested) generated a satisfactory degree of coat-color chimerism. Thus, the original D3 population probably contained only ~20% of germline-competent cells which could efficiently colonize the germline in test chimeras (when 10 - 15 cells per blastocyst were injected). Consequently, 80% of the clones obtained from transfection of this population, which equals to a subcloning step, were unable to contribute to the germline. If you are initiating gene targeting experiments, and consider they will be a major laboratory effort in the future, then we would suggest to first establish (wildtype) subclones of an ES cell line and to test several of these subclones for germline competence. Although this is a significant investment of energy and time, this will most likely circumvent the problem of inefficient ES line germline transmission of a targeted mutation. If a subclone with a good germline record is identified this cell population probably contains almost 100% 'good' ES cells and the majority of derived transfectants should be germline competent for at least for the next 5 - 10 passages of the subclone. Following this strategy we have obtained a good record of germline transmission with an E14 derived subclone (see below).

As mentioned previously, ES cells can be grown either on gelatinized culture plates in the presence of LIF (152, 153) or on feeder layers of primary embryonic fibroblasts (EF; e.g. (143) or from the STO fibroblast cell line (136, 137). STO cells transfected with a LIF expression vector are also available (146). We prefer to culture ES cells on a layer of (G418-resistant) EF cells in the presence of LIF when used for the generation of chimeric mice. If ES cells are just expanded for the preparation of genomic DNA it is sufficient, and even beneficial, to culture them directly on plastic (gelatin-coated) plates in the presence of LIF since the contamination of ES cell DNA with DNA from feeder cells is avoided. Recombinant LIF can be either purchased (Gibco) or prepared from cell lines stably or transiently transfected with LIF expression vectors. We use the supernatant of a stably transfected CHO line (see Appendix) at a dilution of 1:1000 in DMEM-ES as tested by the long term culture of ES cells on gelatinized plates.

The fetal calf serum used for ES cell culture must be tested to support ES cells growth and, upon identification of a good lot, a substantial number of bottles should be reserved as a stock. We test different batches of FCS by determining the plating efficiency and colony morphology after ES cells (grown before with previously tested FCS) are plated out at low density on EF layers in DMEM-ES (with and without LIF) prepared with 15% of the test FCS. ES cells are grown for one passage in the media prepared with the various FCS lots (including the original one) and then 200-500 cells are plated into 6 cm dishes (in duplicate) and colonies are counted after one week. Plating efficiency (number of colonies/number of cells plated) is always well below 100% since some ES cells in suspension, obtained by trypsinization, form small aggregates. Further, the fraction of differentiated colonies is much higher when ES cells are plated out at low density compared to high density (routine) cultures. We regard an FCS batch resulting in ~50% plating efficiency and ~75% of undifferentiated colonies as well suited for ES cell culture. For some sera, results may differ depending on whether the DMEM-ES contains LIF or not. We generally test FCS in media with and without LIF and attempt to identify serum which is able to support ES growth on EF layers without the external addition of LIF.

If the ES cells used are free of mycoplasma (154 for review), then cultures should be regularly checked for their presence since mycoplasma-contaminated ES cells may be unable to generate chimeras and/or to contribute to the germline. Contaminated cultures should be discarded. However, in our experience even a mycoplasma-contaminated ES cell line has demonstrated ability to efficiently colonize the germline (see below). Among the various methods for mycoplasma detection (154) the Gene-Probe system (Gene Probe Inc.), which relies on the detection of

69

mycoplasma rRNA, gives the clearest results in our hands. A PCR-based kit can also be purchased from Stratagene.

ES cells

Currently in our lab there are two ES cell lines which are primarily utilized and a third which is being characterized in our hands:

E14.1 (129-derived; ref. 155)

B6-III (C57BL/6-derived; ref. 150)

BALB/c-I (BALB/C-derived; K. Bürki and B. Ledermann personal communication)

Homologous recombinants have been obtained in all three ES cell lines and all three have been successful in germline transmission, although with different efficiencies (see below). Below are a few brief facts about these different ES cell lines and mice used with each for blastocyst injection and subsequent breeding for germline transmission.

E14.1

This ES cell was subcloned in Cologne from E14 (144, 145) and is derived from the 129/Ola mouse strain (A^w, c^{ch}, p; H-2^b; IgH^a). E14.1 is known to be mycoplasma-contaminated and must always be cultured with penicillin/streptomycin. Nevertheless, the capacity of these cells to transmit to germline is very high. E14.1 cells, like EF cells, tolerate trypsinization (trypsin/EDTA solution in (MT)PBS) quite well. Virtually every E14.1 homologous recombinant generated has been successful in germline transmission. E14.1 has been recloned (E14.1.1) and has tested positive for germline transmission.

In comparison to the other ES cell lines E14.1 appears to be quite robust. That is, it grows very well (doubling time ~24 hours) and can be heavily manipulated without losing its capacity for germline transmission. This is probably best illustrated by the fact that it has been sent to various labs throughout the world and continues to be successful (e.g. (156-158)). Furthermore, the E14.1-derived ES cell line, JHT (58), has been electroporated a total of 4 times with subsequent germline transmission.

Generally, we inject E14.1 ES cells into CB.20 blastocysts (a BALB/c derived recombinant inbred strain (A, c, P; H-2^d; IgH^b) and chimeric animals are seen as a light brown/cream on a white background which is not visible until the pups begin to develop hair (~4-5 days). Mating these chimeric mice to CB.20 results in light brown/yellowish pups with germline transmission and white pups otherwise (from CB.20 x CB.20). Alternatively, injection of E14.1 into C57BL/6 blastocysts (a, C, P; H-2^b; IgH^b) will yield chimeras which can manifest their coat color chimerism in a number of ways such as brown on black or cream on agouti. Mating of chimeras to C57BL/6 animals results in E14 derived agouti colored pups and black pups otherwise. For introduction in coat color genetics see the references (113, 159).

B6III

The B6III ES cell line was derived from a C57BL/6 mouse by Lederman and Bürki (150). These cells are free of mycoplasma. Germline transmission using the B6III ES cell line has been successful although not nearly with the same frequency as with E14.1. The frequency of germline transmission appears to be in the order of 1 out of approximately 3 or 4 clones from clones tested in our lab. Certainly the B6III ES cells are more sensitive than E14 as indicated by the necessity to use the milder modified trypsin solution (see Appendix A). This difference also becomes apparent upon injection. Trypsinization of the B6III ES cells results in a relatively high incidence of cell death and cells which do not look generally healthy. The modified trypsin solution (see Appendix) improves this situation to a degree as does placing the cells at 4°C for a couple of hours.

Chimeras are usually generated in our lab by the injection of these ES cells into CB.20 blastocysts although BALB/c blastocysts can also be used. The resulting chimeric mice are a white background (CB.20 host) with black or brownish (ES cell-derived) chimerism. Good chimerism can often be seen immediately (although not always) in neonates by the presence of black eyes (through the still closed eyelids) versus the reddish/pinkish eyes of the CB.20 host. Breeding of these chimeric mice to C57BL/6 mice result in offspring which are black if ES cell-derived (and genetically C57BL/6 = (BL/6 x BL/6)) or agouti ((BL/6 x CB.20)$_{F1}$) if not ES cell-derived. Alternatively, if crossed to CB.20, germline transmission is indicated by agouti pups ((CB.20 x BL/6)$_{F1}$) or white (CB.20 x CB.20) if not ES cell-derived.

BALB/c-I

The BALB/c-I line (unpublished) has been obtained from B. Ledermann and K. Bürki (Basel) and is derived from the BALB/c strain (A, c; H-2^d; IgH^a). This ES line has been successful in our lab in generating homologous recombinants and the transmission of mutation to the germline although we have not extensively characterized this clone. However, the parental strain is reported to result in chimeras with a high frequency of germline transmission. This ES cell appears to be the most sensitive of all three ES cell lines and the use of the modified trypsin solution (appendix) is mandatory. The mycoplasma status of these cells has not been determined.

Chimeras should be generated using C57BL/6 blastocysts for injection with the resulting chimeras being white on a black background. Germline transmission will be indicated by mating chimeras with C57BL/6 mice and obtaining agouti (BALB/c x BL/6)$_{F1}$ mice vs black mice (BL/6 x BL/6) for nongermline transmission.

All of these ES cell lines are male (XY) and, therefore, germline transmission is usually through male chimeras. However, E14.1 female chimeras have consistently transmitted to germline in our lab and should certainly be bred. It should be noted, however, that when female chimeras transmit to the germline, ~50% of the male offspring are aneuploid and sterile (160). To date, BL6-III female chimeras have not given germline transmission. Furthermore, experience has indicated that germline transmission is usually obtained with higher degree of chimerism (e.g. >60%? by coat color) although exceptions to this have certainly been observed. The evaluation of a targeted clone for germline transmission should be based on a number of criteria which includes the degree of chimerism obtained in chimeric animals and the number of fertile male chimeras obtained (versus sterile). Our rule of thumb is that at least three good, fertile male chimeras should be bred per clone to evaluate the clones capacity to transmit to germline. It should be noted that with BL6-III male chimeras germline transmission has sometimes been obtained only after several (frustrating) months of breeding (and many non-germline litters).

Chapter 15
ES cell culture and transfection

When maintaining ES cells in culture the primary consideration is to prevent their spontaneous differentiation. Any homologous recombinants you obtain will be of no use if the cells are mistreated and lose their ability to colonize the germ cells of the developing chimeric embryo (the exception to this is the possible use of chimeric analysis or blastocyst complementation - see Chapter 21). Therefore great care and consideration should be given to the propagation of these cells in tissue culture. The previous chapter gives a few guidelines used in the routine handling and maintenance of the ES cell lines used in our laboratory.

ES cells used for electroporation should be growing exponentially since the transfection efficiency and perhaps also the frequency at which homologous recombination occurs appears to be increased in optimally growing cells compared to dense ES cultures. For the same reason we prefer to feed ES cultures with fresh medium a few hours prior to transfection. For gene targeting purposes the fraction of cells being in S phase at the point of transfection might be critical as homologous recombination, but not random integration was found to depend on the cell cycle, peaking when DNA is introduced into the nucleus during S phase, as tested in rat-20 cells (161). Thus, ES cells should be plated out in defined numbers 1-2 days before transfection and frequently inspected to meet the point when the cultures contain sufficient cells for transfection but are still in the optimal growth phase.

Prior to transfection the construct must be prepared and linearized. A high quality, clean, large scale preparation of your construct should be made. Commercial columns (e.g. Qiagen™) have been used successfully for preparing the construct for transfection. Before the day of transfection, linearize at least 30 µg of your construct (for every 10^7 cells to be transfected) by restriction enzyme digestion followed by phenol/chloroform and choloroform/isoamyl extraction. Ethanol precipitate your linearized construct and wash twice with 70% EtOH. Air dry the pellet in the open tube in a tissue culture hood. Finally, resuspend in sterile distilled H$_2$O at a final concentration of ~2-3 µg/ml (the point is that you do not want to add your DNA in too large a volume) and quantitate.

For electroporation we usually use a home made electroporator at 240V/480µF and cuvettes from Gibco/BRL but the BioRad electroporation system is equally efficient. Note that BioRad electroporation cuvettes accommodate only 0.8 ml of transfection buffer.

In a typical experiment ~2 x 10^7 ES cells are transfected and ~500-2000 G418^r colonies are expected depending on the neomycin gene and promoter utilized. The synthetic mutant polyoma enhancer/HSV-tk promoter (MC1; (1)) is a weak promoter and generally gives ~500-1000 colonies when used as pMC1neopA. A 3-10x gancyclovir enrichment, therefore, results in ~100-500 double (G418 and gancyclovir)-resistant colonies. The phosphoglycerate kinase promoter (PGK) driving neomycin (and HSV-tk), which is a stronger promoter than the MC1, results in a 4 - 5-fold higher number of G418-resistant colonies but is not necessarily superior to pMC1neopA for gene targeting experiments (see also Section 1, Chapter 3). Besides the neo gene, which is commonly used for the selection of transfectants, the hygromycin B and puromycin resistance genes are established selection markers in ES cells.

In any case, the number of resistant colonies obtained from transfection strongly depends on the concentration of G418 (in the case of the neo gene) used for selection since an increasing fraction of transfectants will not produce sufficient amounts of neomycin phosphotransferase to survive increasing concentrations of G418. This might be related to the influence of the individual integration site on marker gene expression, the number of integrated copies and the promoter strength of the marker gene. Since homologous recombinant ES clones usually contain only one copy of the selectable marker and the potential negative influence of the targeted locus on marker gene expression is unpredictable, the lowest possible G418 concentration should be used for the selection of homologous recombinants. Thus, it is necessary before starting ES cell transfections for gene targeting to determine by titration the lowest concentration of G418 which is just sufficient to kill 100% of nontransfected cells. This concentration varies depending not only on the ES cell line used but also on the neo transgenic mouse strain used for the preparation of EF cells. It must be also taken into account that the available G418 preparations (Gibco/BRL) contain only 50 - 75% of biologically active G418 substance, varying from one batch to another.

ES cell transfection protocol

A schematic representation of transfection is given in Figure 18.

1. Thaw ES cells as quickly as possible (as described above for EF cells) and plate one vial (3-5 x 10^6 cells) onto one plate of mitomycin C-treated EF cells with DMEM-ES medium (see Appendix A). If ES cells are frozen down at higher concentrations then multiple mitomycin C-treated EF plates should be used.

2. Exchange with fresh medium after 24 hours and expand as needed. For general passage, plate ES cells at ~1 x 10^6 cells/9 cm dish on confluent mmc-treated EF cells. Generally, cells must be split every second day.

   E14.1 ES cells double approximately each day therefore if you plate 10^6 cells on day 0 then ~2 x 10^6 on day 1, ~4-5 x 10^6 on day 2, ~1-1.5 x 10^7 on day 3, etc. B6III and BALB/c-I ES cells double somewhat slower than E14.1 although not significantly slower (see Chapter 14 for further details).

   Two days before transfection plate ES cells at higher density than normal at ~2.5-3 x 10^6 cells/9 cm dish x 2 dishes (if you want ≥2 x 10^7 cells for transfection). The number of cells / 9 cm dish should never exceed ~1.5 x 10^7 cells.

   Trypsinization of ES cells should be monitored more carefully than with EF cells. Cells are washed twice with (MT)PBS and 3 ml of trypsin solution added (per 9 cm dish). The E14 ES cells, like EF cells, tolerate the trypsin/EDTA solution (in (MT)PBS; see Chapter 2). The BALB/c-I, as well as the B6III, however, are more sensitive to the trypsin in (MT)PBS resulting in a high degree of cell death. For these lines an alternative trypsin medium is used (see Appendix A).

   Cells are kept at 37°C while trypsinizing until no aggregates are left after resuspending with a pipette (generally 3-5 minutes). You may want to monitor the cells periodically under a microscope determining when they have detached and 'rounded up'.

3. On day of transfection feed ES cells with fresh ES medium a few hours before electroporation, wash with (MT)PBS and trypsinize. Dilute at least two-fold in media, count and aliquot 10^7 cells in 15 ml centrifuge tube. Use 10^7 cells/transfection cuvette/ml of transfection buffer (see Appendix A) and pass or freeze down the rest.

4. Wash cells 1x in (MT)PBS and while cells are spinning down, prepare transfection buffer/DNA mix. Prepare 1 ml transfection buffer with 25-30 µg linearized DNA per 10^7 cells to be transfected. High DNA concentrations during transfection results in higher numbers of stable transfectants but may also be toxic to ES cells when exceeding 50-100 µg DNA / ml transfection buffer.

5. After centrifugation, remove supernatant from pelleted cells and add 1 ml transfection buffer/DNA mix, resuspend well (avoiding bubbles) and transfer to cuvette.

6. Electroporate at 240 V when using cuvettes with 4 mm electrode distance, 1 cm width, and an electroporator with 480 µF capacitance. E14.1, BL/6-III, and BALB/c-I ES cells have all been transfected at this voltage (220-240V).

   After electroporation incubate 5-10 minutes at room temperature

7. Transfer electroporated cells to 30 ml ES media for each 10^7 cells and plate:

10 ml into 9 cm dish x 3 (~3 X 10^6 each; maximum 5 x 10^6 transfected cells/plate)

Control plates

If the transfected cells are to be selected with G418 and GANC (double) selection then:

•10^6 transfected cells are seeded onto another 9 (or 6) cm plate which receives G418 but not GANC to control for transfection efficiency by counting the number of G418-resistant colonies ~10 days after transfection. Comparing the number of G418 (only)-resistant colonies to the number of colonies on the double-selected plates indicates the enrichment from GANC selection.

•5×10^5 *nontransfected* cells are seeded onto a 9 cm dish to control for the stringency of G418 selection. These cells should be fed the same medium as the main part of the transfection (G418 $\pm$ GANC) and should be inspected daily. Selection conditions should be chosen such that nontransfected cells die slowly within ~6 days after the onset of selection. However, no viable wildtype cells should remain on properly selected plates at day 8 of selection when the first transfected colonies may be picked.

•Cell viability after electroporation can be assessed by plating 10^3 transfected and 10^3 nontransfected cells onto two 6 cm dishes. Cells are grown without selection and plating efficiency is determined ~1 week later by counting the number of colonies.

•GANC nonspecific toxicity can be assessed by a control transfection of 2×10^6 cells with a (linearized) vector containing the same neo expression unit as used for the derivation of your targeting vector. These cells are plated onto two 9 cm dishes (10^6 each) one plate receiving medium + G418 and the other medium + G418 **and** GANC. Optimally, the colony number should not differ between plates. In addition, comparing the frequency of G418-resistant colonies obtained from this control transfection with the G418-selected plate of the transfection above (with the targeting vector) will roughly indicate if sequences within the targeting vector suppress the expression of the selectable marker gene.

8. Allow to grow undisturbed for 48 hours (may change medium after 24 hours). At this point, initial selection is started which means feeding ES medium + G418 to transfected cells if the neo gene is used as a selectable marker. Plating efficiency control plate should not be given G418. G418 (or hygromycin, etc., if other selection markers are used) should be given at a tested, minimum concentration just sufficient to kill nontransfected cells.

    G418 concentration: 300 µg/ml for BL/6-III ES cells

    350 µg/ml for E14.1 ES cells

    350 µg/ml for BALB/C-I ES cells

    It should be noted that these concentrations represent total concentrations and were established for G418 with ~50% active concentration.

    Feed cultures daily thereafter for the next 5-6 days with fresh G418-containing ES medium since most nontransfected cells will survive for several days.

9. If thymidine kinase (tk) has been incorporated into the targeting vector for negative selection, then 5 days after transfection (3 days after start of G418 selection) select against random integrants by the addition of GANC to G418 media.

    GANC is prepared fresh each day:

    1) measure 4.3 mg GANC sodium salt (Syntex Cymeven) in Eppendorf tube and add 80 µl ddH$_2$O (= 2×10^{-1} M).

    2) dilute 10 µl of this 2×10^{-1} M in 1 ml ES medium (= 2×10^{-3} M).

    3) filter sterilize using small millipore syringe filter and dilute 10 µl of this 2×10^{-3} M in 10 ml ES medium-containing G418 giving a final concentration of 2×10^{-6} M (2 µM GANC).

    G418 control should not be given GANC.

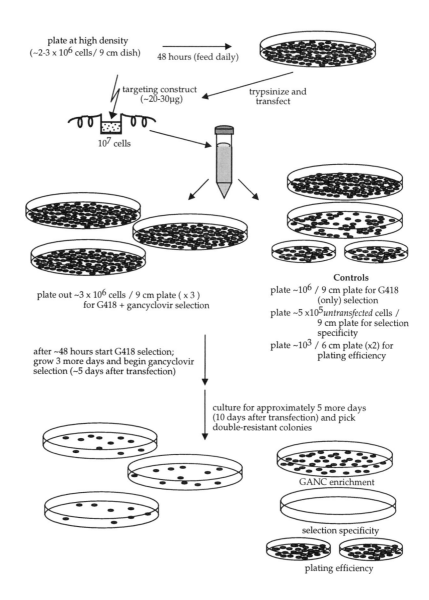

plate at high density
(~2-3 x 10^6 cells/ 9 cm dish)

48 hours (feed daily)

targeting construct
(~20-30µg)

trypsinize and
transfect

10^7 cells

plate out ~3 x 10^6 cells / 9 cm plate (x 3)
for G418 + gancyclovir selection

Controls
plate ~10^6 / 9 cm plate for G418
(only) selection

plate ~5 x10^5 *untransfected* cells /
9 cm plate for selection
specificity

plate ~10^3 / 6 cm plate (x2) for
plating efficiency

after ~48 hours start G418 selection;
grow 3 more days and begin gancyclovir
selection (~5 days after transfection)

culture for approximately 5 more days
(10 days after transfection) and pick
double-resistant colonies

GANC enrichment

selection specificity

plating efficiency

Figure 18. Schematic overview of ES cell transfection including the control plates
which must be considered. See text for details.

Chapter 16
ES clone picking

The number of clones that must be recovered will depend on a number of factors such as the type of selection (e.g. selection for homologous recombinants and against random integrants), number of colonies obtained after selection, method of screening for homologous recombinants (Southern vs PCR), knowledge about the targeting frequency for the locus of interest, etc. For obvious reasons, however, the more clones isolated the higher the probability of recovering a homologous recombinant. In situations in which we have targeted the same locus in multiple instances (e.g. the immunoglobulin heavy chain locus) we have generally observed a similar frequency of homologous recombination events and therefore know approximately how many colonies must be isolated to obtain a homologous recombinant. As a very general number, at least 250-300 colonies should be isolated for examination of a homologous recombination event.

1. Once selection is complete (~10 days after transfection for G418 and gancyclovir selection), resistant clones must be isolated. Colonies at this point should be fairly large in size (~4000 cells) but still maintaining a discrete border or edge. Colonies may even have a dark 'cap' indicative of a high density of cells at the top of the colony. What is to be avoided are the colonies which have begun to differentiate (and spread out) around the edges. Concerning the size of resistant colonies at a given time, there is a considerable variation among the population of resistant clones obtained by transfection. Thus, it is probably best to pick every day, starting at ~day 10 after transfection, the colonies which have grown to the proper size. After one week (~day 17 after transfection), however, almost all colonies are overgrown and the plates should be discarded. One should be cautious when picking colonies too early after selection. These colonies, while clearly visible, are often small and make it difficult to evaluate their state of differentiation.

2. One day before picking, prepare enough mitomycin C-treated 48 well plates to accommodate all clones (2x if screening by Southern). Plate ~3.5-4 x 10^6 EF cells/48 well plate in 0.5 ml/well resulting in ~ 8 x 10^5 cells/well.

3. Add 1x trypsin/EDTA (50 μl/well) to 96 well plate and 0.5 ml ES medium to mitomycin C-treated 48 well plates. U-bottom (or V-bottom) 96 well plates are preferable to flat bottom since this will make recovery much easier. Trypsin should not be added to 96 well plate too far in advance since significant evaporation can occur by the time colonies are added. Mark rows at this point as well (see below).

4. Wash ES cells 1x in (MT)PBS and cover cells with >12 ml (MT)PBS. (MT)PBS should not be too warm as the feeder layer may disassociate from the plate too quickly.

5. With a dissecting scope in a tissue culture hood (with outward pressure) and using a cotton-plugged pasteur pipet with modified bulb (see below), gently dislodge a colony from the feeder layer and withdraw it in a minimal amount of (MT)PBS. Transfer immediately to 96 well plate containing trypsin. Continue until finished or until feeder layer begins to come up but perhaps no longer than 30 minutes. If you are not picking all colonies at once, ES medium may be added and cells placed back in the incubator. While picking colonies, the 96 well plate can be examined under an appropriate microscope to insure that each well contains a picked colony.

The bulb used at the end of the pasteur pipette is actually a small piece of flexible tubing which has been plugged on one end. The advantage to this is that a normal rubber bulb has too much resistance and will very quickly induce fatigue in the fingers which becomes significant when picking several hundred colonies. There are several different techniques for the actual picking of ES colonies which differ primarily in the tool used to remove the colonies. The method of choice is largely a one of personal preference. Alternative methods for picking colonies include using a mouth pipette attached to a plugged (drawn out) pasteur pipette or a P-200 with sterile yellow tips although the disadvantage to the latter method is the inability of seeing the colony being taken up by the pipette tip.

As you are picking colonies it is easy to confuse which colony corresponds to which well of the plate and may result in mixing colonies. Similarly, transferring the trypsinized colonies from the 96 well plate to the 48 well plate may also be a source of mixing up colonies and/or wells. Therefore some methodology should be established to avoid this type of confusion. For example, a good idea is to add colonies to the 96 well plate in rows of 8 (vs rows of 12) and by using only every other row marking each row with a marker or placing each used pasteur pipette (or yellow tip) on the table and removing all eight after a row is finished (or doing both). The marked rows of 8 in the 96 well plates will also help in transferring the trypsinized colonies to the 48 well plate since these are also in rows of 8 (Figure 19).

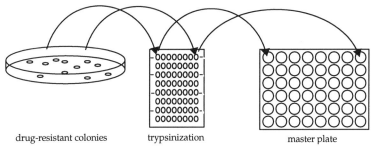

drug-resistant colonies   trypsinization   master plate

Figure 19. To avoid mixing colonies, a method should be considered for the transfer of colonies from the dish to the 96 well plate and, after trypsinization, to the 48 well plate.

6. From the 96 well plate, pipette each well ~5 times and transfer half volume to 48 well master plate. This is perhaps best accomplished by adding 50 μl of ES medium from the corresponding well of the 48 well plate, resuspend cells by pipetting, and transferring 50 μl back to the master plate. For PCR screening, transfer second half to PCR tube for amplification or, for Southern screening, transfer second half to a duplicate 48 well plate for expansion. Depending on the number of colonies to screen, clones can be pooled for screening by PCR. Once master clones have grown sufficiently they can be frozen down directly in 48 well plate.

7. For screening by Southern analysis, after sufficient growth (usually several days) trypsinize duplicate plate in 100 μl trypsin and transfer to larger plate (e.g. 12 or 6 well plate depending on cell density). If cells need to be expanded only to isolate genomic DNA, they then can be transferred directly to plastic (gelatin-treated) plates and do not need to be plated on mmc-treated EF cells. If a Southern strategy is not optimized with respect to probe and enzyme, it is advisable to grow clones up to a sufficient cell number in order to obtain enough DNA for several Southerns. From a dense single well of a 6 well plate enough DNA is obtained for multiple digestions (~2-3).

When gelatin-treated plates are used, tissue culture dishes are first incubate with sterile, autoclaved 0.1% gelatin (in (MT)PBS or H_2O) in enough volume to cover the plate for at least 0.5-1 hour at room temperature in the hood. Gelatin used for this is from porcine skin / 300 bloom (Sigma no. G-1890). Aspirate gelatin and add ES cells in ES-medium directly to plate.

81

Chapter 17
Identification of homologous recombinants

One of the biggest strategical questions you must ask about your targeting experiment (preferably before vector construction) is whether or not to screen by PCR. A major advantage to screening by PCR is that the number of colonies which can feasibly be analyzed is much larger than that for screening by Southern analysis. This will of course be important if the locus you wish to target has a low efficiency of undergoing homologous recombination (for unknown reason(s)) and may make the difference between finding a homologous recombinant or not. Another advantage to PCR screening is being able to very quickly discard clones which have not homologously recombined, greatly reducing the amount of tissue culture required when screening by Southern analysis.

If you have finished the construction of your targeting vector before considering how to screen for recombinants, then PCR screening will not necessarily expedite the identification of homologous recombinants compared with Southern analysis. This is because proper PCR conditions should be established using ES cells transfected with a PCR positive control construct prior to transfection of the targeting construct (PCR conditions with purified DNA ≠ PCR conditions with ES cells). Failure to do so has often resulted in mad, frantic scrambles to optimize, or even obtain, PCR conditions just prior to picking colonies (not a nice scenario). PCR screening will be faster, however, if the decision to screen by PCR has been made early (e.g. prior to or during the construction of the target construct) and conditions already established on a positive control ES cell clone (see below). Nevertheless, homologous recombinants identified by PCR are only considered putative until confirmed by Southern analysis. Therefore, irrespective of your method for screening colonies a strategy for Southern analysis must be developed. This means an appropriate enzyme and probe must be identified and demonstrated to be valid for the purposes of screening. For a large number of colonies screening by

Southern becomes cumbersome and tedious as you must handle (feed and subsequently freeze) all of the clones while recombinants are identified.

In the end, the decision is largely a personal one depending often on many factors such as availability of genomic sequence to make a PCR control construct, the design of the targeting construct (e.g. no short arm of homology), the number of colonies obtained after selection, etc.

Finally, whichever method you choose for screening your recombinants, it is highly recommended (essential?) that the method is shown to be reliable **before** actual transfection. For PCR screening this means that positive control ES cells are generated and conditions optimized. For Southern screening this means probe(s) are identified and already known to give a good signal (and not just possession of a genomic clone encompassing a potential probe). Restriction enzyme(s) should be chosen which will clearly discriminate between random and targeted integrations.

Screening by PCR

This involves the generation of a control construct and optimizing the PCR conditions for the sequence which will indicate a homologous recombination event. The essential features of a positive control construct are that it contains the short arm of homology, a primer sequence found within the construct and not in the endogenous genomic locus (e.g. neomycin, or a selectable marker, is generally used for this) **and** sequence found <u>outside</u> the construct but within the endogenous locus (Figure 20). This construct is often the same as the targeting construct (although does not necessarily need to be) but with an extension of the short arm of homology.

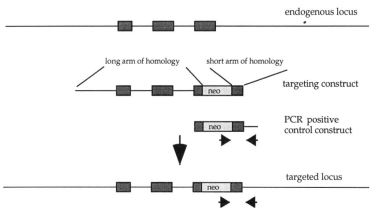

Figure 20. PCR strategy for detection of homologous recombinants. A primer pair (small arrows) is chosen such that one primer anneals to a sequence specific to the targeting construct (and not endogenous locus) and the other primer anneals to a region outside the sequence used in the targeting construct. A control construct is usually necessary for determining amplification conditions (see text).

83

Once a positive control construct is made it is transfected into ES cells and transfectants identified which harbor a randomly integrated, single copy of the construct. This is important because you want to establish the PCR conditions on cells which most closely mimic the true situation and that means a single targeted locus. With this positive control clone the PCR conditions are established. Initially we determine the PCR conditions which give the highest signal to noise ratio. Generally, this means starting at an annealing temperature below that of the lowest primer Tm and varying the magnesium concentration. Once an optimal magnesium concentration is identified, then use this concentration and vary the annealing temperature by increasing it until a strong, specific signal is still obtained. Once these conditions are determined the next step is to dilute the positive control ES cells with wildtype ES cells. The rationale for this is that you will more than likely be pooling colonies for screening and you must be able to detect a single targeted locus among an excess of wildtype loci.

ES cell PCR amplification on positive control clones (162).

1. Aliquot ~10^4 positive control ES cells in a PCR tube, wash with (MT)PBS, and aspirate supernatant leaving ~5 µl of (MT)PBS (may also freeze these aliquots at -20°C). Be careful not to aspirate cell pellet; a drawn out Pasteur pipet facilitates removing the supernatant.

2. Add 20 µl of 1.25x Taq polymerase buffer (1x final), overlay with mineral oil, and incubate at 95°C for 10 minutes.

3. Add 10 µg proteinase K from 10 mg/ml stock. Important to physically mix the proteinase K and cells.

4. Incubate at 55°C for 1 hour followed by 95°C for 10 minutes (to inactivate the proteinase K).

5. Add 25 µl of 2x concentrated PCR mix:

   5 µl dNTPs (at 2 mM)   = 200 µM final concentration

   1 µl of each primer (at 25 µM)   = 0.5 µM final concentration

   0.5 µl Taq polymerase (5 U/ml) = 2.5 U/reaction in 1x Taq buffer.

6. Amplify according to conditions necessary for your sequence. Start by optimizing Mg^{++} concentration and then annealing temperature.

7. Once optimal conditions are established, dilute positive control ES cells with wildtype ES cells to determine the sensitivity of your PCR. Dilutions should be in the order of 1:5 - 1:500, always maintaining ~10^4 cells in a reaction mix.

Screening by Southern analysis

1. Identify restriction enzyme(s) and probe(s). The results of your Southern analysis should be unambiguous since it is essentially based on this result that you proceed with the generation of a targeted mouse. Accordingly, the choice of enzyme and probe is important in verifying that a homologous recombination event has occurred. Ultimately, the Southern should demonstrate that a homologous recombination event has occurred at the short arm of homology, the long arm of homology, and that no gene duplication events have occurred during the recombination.

2. Ideally, probes are used which do not hybridize (entirely) within the targeting construct, i.e., come from just outside the construct. The criteria for which restriction enzyme to use is that it should generate a fragment unique to the homologous recombination event and which will be labeled by the probe. For example, in Figure 21 with probe **A** just outside the short arm of homology, the enzyme should cut within the targeting construct at a unique introduced site (e.g. enzyme S) or on the opposite side of the selectable marker and outside the construct and beyond the probe (e.g. enzyme R). With the former enzyme (S), probe **A** will detect a fragment smaller than the endogenous wild type fragment and with the latter enzyme (R) a band ~1.2 kb larger (equal to the size of the introduced selection marker) than the endogenous fragment. A similar strategy is used for verifying the integrity of the long arm of homology (probe **C** and enzyme S). For examining gene duplication a probe entirely internal (and possibly unique to the targeting construct, e.g. probe **B**) of the targeting construct is used.

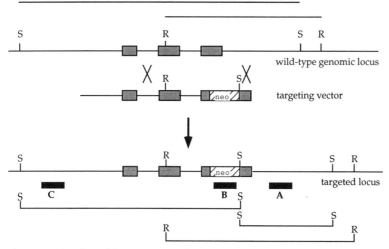

Figure 21. Southern blot strategy. Choice of probes and restriction digests are critical for the verification of a homologous recombinant event. Importantly, your strategy should be able to distinguish between duplications or deletions as well.

Chapter 18
Freezing ES cell clones

One of the biggest advantages to screening your colonies by PCR is that you can immediately focus on a smaller, more manageable number of colonies. This is because potential homologous recombinants are identified essentially the same day as the colonies are picked. In practical terms this abrogates the need to freeze **all** of your picked colonies (often ≥ 300 colonies) which is a large task. Nevertheless, even with PCR screening the time comes when you (may) need to freeze down ES clones to verify the targeting event by Southern analysis. Depending on the number of potential colonies you need to freeze down (and when screening by Southern analysis, all the colonies are potential recombinants) it is often convenient to freeze down these colonies in 48 well plates. There are different methods to accomplish this which depend on whether you freeze down trypsinized cells or whether you trypsinize after thawing. The former method is given here.

The objective of freezing down ES cells is to be able to subsequently recover the desired clones. A rule of thumb is that there should be enough cells within each well such that if only 10% are viable after thawing you will still have enough cells to recover the clone. The other side of this is that a well too dense with colonies is not good and will often lead to differentiation. The point at which to freeze down the plate can also be a problem when the colonies picked have large differences in colony size resulting in different densities between wells of the same 48 well plate. If this is the case, then obviously freezing down the plate at a certain time will not be optimal for all clones and some kind of compromise must be made. (It is possible to selectively freeze down wells but, again, depending on the number of colonies this may or may not be practical.)

If cells are to be frozen in conventional cryotubes, trypsinize cells, spin down and resuspend such that at least 10^6 cells are contained in 0.5 - 1 ml of freezing medium (10% DMSO/90% DMEM-ES cell media or 90% FCS; DMSO should be kept at -20°C in aliquots) per vial. Place vials in cardboard box or styrofoam container to allow slow freezing at -70°C.

Cells can be kept at -70°C for several months but are better transferred into liquid nitrogen for long term storage.

For 48 well plates:

1. Trypsinize cells with 100 µl/well of 1x trypsin at 37° C for ~3 minutes (or as long as necessary to transfer freezing medium to wells - see below).

2. During trypsinization aliquot 0.4 ml freezing medium/well to a 48 well plate. This is not done earlier so as to not change the pH of the freezing medium too much (or, alternatively, HEPES may be added to buffer against pH fluctuation). Freezing medium should be initially at 12.5% DMSO (e.g. 3 ml DMSO + 21 ml ES medium for one 48 well plate) so that after addition of 100 µl trypsinized cells the final concentration is 10% DMSO.

3. Disperse cells with pipetman and transfer cells into 48 well plate containing 0.4 ml freezing medium (final concentration = 10% DMSO).

4. Mix gently, wrap in saran wrap, and place plate in cardboard freezing box at -80° C. Clones have been recoverable when stored in this way for several months but should not be in a freezer which is often accessed to prevent temperature fluctuations.

Freezing medium:

5 ml DMSO + 35 ml ES medium = 12.5% DMSO

or 1 ml DMSO/8 ml ES medium

Chapter 19
Thawing ES clones from 48 well plates

After identification of homologous recombinants by Southern analysis, plates must be thawed and appropriate clones recovered. Again there are different ways to thaw the frozen plates which differ slightly in the time required for the thaw.

1. With either method beforehand prepare a 15 ml tube with 10 ml warm ES-media for every clone to recover and leave in hood.

2. Thaw the plate at 37°C. The safest, but slightly longer, method is to place the frozen plate in the 37°C incubator. Alternatively, the frozen plate is wrapped in new saran wrap and, holding the excess wrap above the plate, submerge/agitate the plate-bottom in a 37°C water bath. Of course with the latter method it is imperative that no water can touch the plate while thawing. With either method you will observe that the outer wells will thaw first and the plate should be taken to the hood before the inner wells are completely thawed.

3. At this point take the plate to the hood and begin to transfer the contents of each well to a different 15 ml tube.

4. Resuspend pellet with fresh DMEM-ES and transfer to an appropriate sized (depending on cell number) mmc-treated EF plate.

Chapter 20
Genomic DNA preparation

There are several methods for extracting genomic DNA from tissue culture cells. Presented here are two methods: a relatively new method which is very simple, quick and convenient (no phenol) and a reliable, although perhaps outdated (i.e., phenol/chloroform), method. Genomic DNA can be extracted from either tissue culture cells (or mouse tails) using either method. The simplified, non-phenol method presented here was developed with gene targeting in mind since large numbers of genomic extractions are required from both tissue culture cells and mouse tails in the course of an experiment (163). Genomic DNA isolated from ES cells and tails in this manner has been used successfully in Southern analysis and PCR amplification.

Non-phenol method

Tissue culture DNA isolation

The original protocol indicates cells can be lysed directly in the well although an initial trypsinization and wash will expedite the procedure and insure good recovery of DNA. Furthermore, if cells have been grown on gelatin, then lysing cells directly in the well with lysis buffer results in the gelatin forming an insoluble mass making the DNA difficult to separate from the solution. Growing cells for genomic DNA preparation directly on plastic will prevent this problem.

For up to a 6 cm dish :

1. Trypsinize, transfer to an Eppendorf and wash cells 1x with (MT)PBS. Resuspend in 500 µl of lysis buffer.

   Lysis buffer: 100 mM Tris-Cl pH ~8.5

   5 mM EDTA

> 0.2% SDS
>
> 200 mM NaCl
>
> 100 µg/ml Proteinase K

2. Incubate at 37°C for several hours or until homogenous (no cellular debris). Agitate periodically during incubation.

3. Add equal volume of room temperature isopropanol and mix well by inversion. DNA precipitate should be visible.

4. Recover DNA precipitate by using a drawn out pasteur pipet and lifting DNA out of the alcohol (spinning out). Dip DNA into 70% EtOH and transfer to new Eppendorf tube. Allow to briefly air dry and resuspend in TE. Amount of TE depends on the size of the precipitate but is generally 50-150 µl. Alternatively, if PCR is to be performed with the DNA, or with very low yields, the precipitate can be centrifuged in a microfuge, washed with 70% EtOH and resuspended in TE. This will generally result in more difficulty in resuspending the DNA (e.g may need to incubate at 37°C to fully resuspend the DNA).

Tail DNA isolation

1. Cut ~1 cm tail into 1.5 ml Eppendorf tube.

2. Add 700 µl lysis buffer (100 mM Tris-Cl pH ~8.5, 5 mM EDTA, 0.2% SDS, 200 mM NaCl, 100-400 µg/ml proteinase K).

3. Incubate at 55°C several hours to overnight (or until tissue is dissolved) agitating as often as possible.

4. Spin at max. speed for 10 minutes to pellet hair and debris.

5. Pour off supernatant directly into equal volume isopropanol.

6. Mix and spin out genomic DNA, immerse in 70% EtOH and transfer to a new tube. Resuspend DNA in ~150-200 µl TE. Alternatively, spin down genomic DNA and wash 1x with 70% EtOH, air dry briefly and resuspend pellet in TE. Spinning down DNA may necessitate incubation at 37-55°C for several hours and vortexing to resuspend the DNA. Some insoluble material often remains but does not appear to interfere with enzyme digestions.

7. Use 1/5 to 1/10 of total volume for digestion.

Phenol/chloroform method

Tissue culture DNA isolation

1. Trypsinize and wash cells 1x with TSE. Resuspend pellet well in 250 µl TSE (TSE = 10 mM Tris, 150 mM NaCl, 10 mM EDTA).

2. Add 250 µl of TSE containing 0.4% SDS and 0.6-0.8 mg/ml proteinase K (final concentration 0.2% SDS and 0.3-0.4 mg/ml proteinase K). The point here is that if you try and resuspend the cell pellet in the final solution of SDS/proteinase K then you will end up with an insoluble mess. Therefore, resuspend first in TSE alone and then add 2x SDS/proteinase K.

3. Incubate 55°C overnight (at least 5-6 hours) or until no cellular debris is visible.

4. Phenol/chloroform extract 2x and chloroform/isoamyl alcohol (24:1) extract once.

5. Ethanol precipitate. At this point precipitated genomic DNA should be visible. If not, spin down in a microfuge for 10 minutes, wash once with 70% EtOH, and resuspend in minimal volume of TE. Alternatively, if precipitate is visible, draw out a pasteur pipet with flame (and seal) and use to spin out DNA precipitate. Immerse once in 70% EtOH and transfer to new tube. After removal of EtOH traces may resuspend immediately in TE. The advantage to spinning the DNA out is that it can be resuspended almost immediately and digestions can be done right away.

Tail DNA isolation

1. Cut ~1 cm tail into 1.5 ml Eppendorf tube.

2. Add 700 µl lysis buffer (50 mM Tris-Cl pH 8, 100 mM EDTA, 100 mM NaCl, 1% SDS, 100 - 400 µg proteinase K/ml).

3. Incubate several hours to overnight at 55°C with occasional agitation until tissue dissolved.

4. Phenol/chloroform (500 µl) extract 1x or 2x (10 second vortexing) and chloroform/isoamyl alcohol (24:1) extract once.

5. Add 500 µl isopropanol and mix until precipitate forms. Spin for 30 seconds, remove supernatant and wash pellet once with 500 µl 70% ethanol. Carefully remove the supernatant and air dry.

6. Add 150 µl TE and incubate several hours at 37°C or overnight at room temperature. Mix well. Use ~ 25 µl (~6 µg DNA) for digestion.

91

Chapter 21
Cre-mediated (neomycin) deletion in ES cells

As discussed in Section 1, a major advantage to using the Cre/loxP recombinase system in conjunction with gene targeting is the ability to remove the floxed selection marker (as well as any additional loxP-flanked sequences) after homologous recombinants have been identified. This is clearly important when studying gene regulation to avoid the influence of the selection marker, and its own regulatory elements, on the regulation of the gene of interest (20). Removing the selection marker (e.g neo gene) from homologous recombinant clones also allows for the selection of ES cells which are homozygous or hemizygous (if Cre-mediated deletion deletes the allele) for the desired genetic modification after retransfecting these clones with the initial targeting vector.

In the absence, or instead, of germline transmission, homozygous mutant ES cells can be used to analyze the introduced mutation either in fetal development or in a lineage specific manner (164, 165). Nagy and Rossant and colleagues have demonstrated that ES cells will direct the development of the fetus when aggregated together with tetraploid mouse embryos resulting in a fetus which is completely ES cell-derived (147, 166). Thus, this technique permits mutations introduced in ES cells to be rapidly assessed in fetal development and, in particular, when such perturbations are embryonic lethal (167). Chen and colleagues (164) have pioneered the use of 'blastocyst complementation' in which ES cell-derived lineages can be directly examined in chimeric mice when the host blastocysts harbor a lineage-specific cell autonomous defect. In this case, host blastocysts deficient in an enzyme which is critical for the development of both T and B lymphocytes (recombinase activating gene - RAG1/2) are injected with homozygous mutant ES cells. Any lymphocytes which arise in resulting chimeric animals are ES cell-derived and can be assessed for the mutation.

Although homozygous mutant ES cells can be generated by alternate means, such as using two identical targeting vectors with different selection markers or increasing the concentration of selection drug, the use of Cre/loxP for this purpose, in our hands, is less cumbersome than the former and more reliable than the latter method. Furthermore, the use of different selection markers also necessitates the preparation and use of EF cells resistant for both selecting drugs.

If your final objective is simply the generation of a mouse strain devoid of the selection marker (or other floxed sequences), then Cre-mediated deletion in ES cells is not necessarily required. In this case, the floxed allele is first transmitted through the germline and resulting mutant mice are subsequently bred to a Cre transgenic which expresses the Cre recombinase early in embryonic development (96, 97). The progeny of such crosses harbor mutant alleles which have undergone loxP-specific recombination in all tissues, including the germ cells. This method of removing the selection marker abrogates the need for an additional electroporation step (for transient Cre expression) and, importantly, reduces the time ES cells are maintained in culture.

To delete floxed sequences in vitro, the Cre recombinase is transiently expressed in ES cells and clones which have lost the neomycin gene by Cre-mediated, loxP-specific recombination are isolated as initially described for ES cells by Gu *et al.* (58). If a selectable marker, such as thymidine kinase, has been included within the floxed sequence, then the isolation of neomycin deleted clones is straightforward. However, thymidine kinase is often used for the selection of homologous recombinants and is, therefore, not available for the selection of Cre-mediated neomycin deletion. In these instances neo-deleted clones must be identified on the basis of G418 sensitivity. Below is a protocol for the isolation of clones which have deleted a floxed neomycin gene and is shown schematically in Figure 22. A more detailed discussion and consideration of the Cre/loxP recombination system is provided in Section 1.

1. Transfect 0.5-1×10^7 homologous recombinant ES cells with 5-30 μg of supercoiled Cre recombinase vector. This number of transfected cells is in vast excess of what will actually be screened for deletion (see below) but the remaining cells can be frozen down for a later analysis if the first is unsuccessful. The amount of Cre recombinase vector used for transfection may vary depending on type of deletion desired although does not necessarily always exhibit a good correlation with the type of deletion when multiple loxP sites are present. This means that use of much less recombinase vector for transfection will not necessarily ensure a type II deletion (see Figure 9). Presently, the two Cre expression vectors generated by Hua Gu (pIC-Cre and pMC-Cre - see Appendix C) have been used successfully for deletion in ES cells. The difference between the two is that the latter Cre contains a nuclear localization signal although for the purpose of transient expression in ES cells either Cre vector is sufficient. Use transfection conditions as

93

established for the ES cells in original transfection of targeting construct.

2. Plate out transfected cells (on mitomycin C-treated EF cells) at ~ 5 x 10^6 cells/9 cm dish and allow to grow for ~48 hours with normal ES medium. This initial plating is to allow the Cre recombinase to be expressed and to avoid the situation where Cre deletion occurs as a colony is expanding resulting in a colony which contains both deleted and undeleted ES cells.

3a. After 2 days (or as soon as colonies are established and Cre has had an opportunity to delete) trypsinize plates and replate (on mitomycin C-treated EF cells) at ~10^3 cells/9 cm plate. Freeze remaining transfectants for subsequent plating and picking, if required. The number of plates depends on number of colonies you wish to screen (and, therefore, efficiency of transfection and Cre-mediated deletion). In general, one plate should suffice to pick 100 colonies which should include 1-10 deleted clones although you may want to plate out several plates in order to divide the colony picking between different plates. In our experience the deletion of a loxP flanked neo gene occurs at the very minimum 50% of the frequency of transient transfection (e.g. 5% deleted clones when 10% of cells are transfected). The efficiency of transient transfection may be controlled by transfection with a β-galactosidase expression vector (e.g. pCH110; for protocol see Appendix C).

3b. In general, we have abandoned using tk for positively enriching clones that have undergone Cre-mediated neo deletion since this is generally an efficient event. When it is used, however, the protocol should be to plate out Cre transfected cells at a lower density (~0.1-1 x 10^6 cells versus 5 x 10^6/9 cm dish) and then to start selection 48 hours post transfection on this initial plate. The idea is to plate out enough cells to obtain clones that have deleted the selection markers but not so many that it will be hard to pick individual colonies. A few different densities should be plated.

4. Picked colonies should be trypsinized in 50 μl in a 96 well plate (V- or U-bottom plates are preferred) as in original colony picking and transferred to a master and duplicate 48 well plate containing mitomycin C-treated EF cells.

5. Allow colonies to become established (again, ~48-72 hours) and add G418 to duplicate plate to test for G418-sensitivity (neo deletion). Amount of G418 added should be at least the same concentration of G418 as used in original selection although using ~25% more G418 may make it easier to detect sensitive (dying) colonies. Dying colonies should be visible after 24-72 hours and after 5 days G418-sensitive colonies should be unambiguous. Sensitive clones should not have **any** viable colonies remaining in well.

6. Clones identified as G418-sensitive should be expanded (from duplicate master plate) for freezing, to retest for G418 sensitivity, and genomic DNA isolation to verify deletion event by Southern. It is important to avoid a mixed colony which may also contain non-deleted cells and/or wild type ES cells (which would be G418-sensitive).

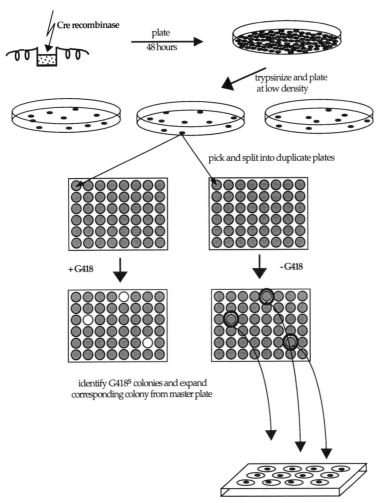

Figure 22. Scheme for the identification of ES clones which are drug-sensitive due to Cre-mediated deletion of a floxed selection marker

95

Chapter 22
Generation of mice - preparation

There are several methods in the literature which result in ES cell-directed development and germline colonization of a developing mouse blastocyst (16, 140, 147, 168-170). We describe here two methods: the physical injection of ES cells into the blastocoel of blastocysts and the aggregation of ES cells with eight-cell stage/morulas. Although perhaps not as useful for most purposes, ES cells have also been shown to be able to direct the development of a fetus when aggregated together with tetraploid embryos, possibly resulting in viable offspring (147). In our view, the decision as to which of the two general methods to use depends on personal preference but may also be determined by the equipment available, the number of chimeras that need to be produced, the ES cell line used, and the individual's skill in micromanipulation. In contrast to blastocyst injection, aggregation methods require no expensive equipment (microscope, micromanipulators) and allow the processing of larger numbers of embryos per day. For the production of low numbers of chimeras for transmission of a targeted mutation through the germline, however, both methods are equally useful.

Finally, as the production of chimeras is animal experimentation, make sure that you are working according to institutional or otherwise appropriate guidelines.

Preparation

You finally have homologous recombinants, have successfully deleted loxP flanked sequences, and are now ready to generate mice with your correctly targeted ES cells. The first thing to think about is mice breeding. This is because if you wish to begin your blastocyst injections on a Monday, for example, this requires setting up mice for blastocysts on the previous Thursday and mice for foster mothers on Friday (explained in detail below).

Your ES cells must also be considered since they should be at an appropriate density for blastocyst injection or embryo aggregation. Typically, at least two days prior to injection (aggregation) they should be plated onto feeders (without feeders for aggregation) in a 12 well dish at varying densities (~0.1 - 1 x 10^5 / well). This allows you to examine the different densities/sizes of colonies and choose the wells which look the best for your injection or aggregation. Furthermore, when injecting, several wells of each density should be available so multiple trypsinizations can be performed throughout the injection day on essentially the same cells.

If you will inject ES cells into blastocysts, then needles must also be made prior to the start of injection. This means transfer, holding, and injection needles. It is a good idea not to wait until the last minute to make your needles and to have a large supply of good backups. Frustration with injection often reaches its peak at a time when a needle breaks and you are out of new needles to use while your ES cells and blastocysts grow old in your injection chamber.

Finally, the anesthesia (see Appendix A) should be prepared and, optimally, tested before the transfer surgery. For an introduction to the anatomy of the male and female reproductive tract, mating behavior and early embryonic development we refer you to the references (171, 172).

97

Chapter 23
Mouse breeding for the generation of ES cell chimeras

When ready for blastocyst injection or aggregation, mice breeding must be set up for two purposes:

1. A source of blastocysts for ES cell injection or aggregation.

2. Foster mothers as hosts for injected or aggregated blastocysts.

Breeding for donor embryos

For ES cell injection blastocysts are needed at day 3.5 after fertilization therefore breeding must be started one day earlier, or 4.5 days prior to injection date. For example, to start injections on a Monday mice breeding must be set up on Thursday of the preceding week. Embryos for embryo aggregation are used one day earlier, at day 2.5 post-coitus (p.c.). For normal matings, two females of breeding age (≥8 weeks old) are placed together with a single stud male (also of breeding age) in a small cage. The following morning females are checked for a vaginal plug which is indicative that a mating has occurred. These plug-positive females are separated and that day is considered day 0 (assuming mating occurred after 12 midnight). These plug-positive females must be replaced by new females and plug checks continued on the subsequent mornings for as long as required for ES cell injections. If you are breeding longer than one week it is best to use a second set of breeder males for the following week since the plug rate decreases after a few days if females are continually housed together with males.

For the injection of 25-50 blastocysts a minimum of 15-20 cages should generally be set up although many factors influence this number such as mouse strain (CB.20 provides more blastocysts/plug positive female than C57BL/6 e.g. 5 vs 3 in our facilities), health of animals, condition of the animal facility (i.e., stress), superovulation (see below), time of year, etc. Female mice ovulate every four to five days so that every day of mating, if the males perform well, 20 - 25% of the females should be plug positive on average. However, often there is a considerable day-to-day variation in the number of plugged F_1 females. Since the ovulation of females within one cage tends to be synchronized it is best to use females for mating which were stored in different cages. Ultimately, you will adjust the cage number as required after the first day or two of plug checks after determining the number of plug-positive females being generated per day.

In some instances superovulation may be necessary to enhance the number of usable embryos/female, for example with some mutant strains (e.g. RAG-2$^{-/-}$) or BL/6 mice. Superovulation can be induced by hormonal i.p. injections of 5 U pregnant mare serum gonadotropin (PMSG; mimics follicle stimulating hormone) in the morning or afternoon followed 46-48 hours later with 5 IU human chorionic gonadotropin (hCG; mimics luteinizing hormone) into sexual immature, young females (see Appendix A for hormone preparation). After administration of hCG these females are housed overnight with males. The females are checked for plugs and caged separately the next morning. Males should be used every second day for mating. If C57BL/6 females are used, for example, often only ~50% of the hormonally treated females are plug positive. Furthermore, these females are variable with respect to embryo production with some containing up to 20 blastocysts at day 3.5 p.c. and others with no blastocysts. Unfortunately, the success of superovulation is quite variable depending on the mouse strain, age and weight of females (~3-4 weeks best for C57BL/6; (173), 8 weeks best for CD1 mice) and preparation and dose of hormone. Since it is usually difficult to keep suitable conditions constant, we generally rely on normal mating for blastocyst production but eventually use superovulation when larger numbers of morulas are needed for embryo aggregation.

Finally the number of embryos to be used or aggregated must be considered. Ideally one should expect ~15-20% of blastocysts transferred into foster mothers to develop into pups of which 50-90% should be chimeric. If E14.1 derived ES cells are injected not more than 50 blastocysts/morulas should be transferred per ES clone since this number is usually sufficient to generate at least one germline chimera. For germline transmission of a targeted mutation we generally use two independent recombinant clones (~50 blastocysts per clone) for chimera production.

Breeding for embryo host (foster mother)

Generally F_1 females, e.g. (C57BL/6 x BALB/c)F_1 or (C57BL/6 x CBA)F_1 mice, are used as foster mothers as they exhibit good maternal instincts (although the occasional pups are neglected or eaten) and reproduce fairly well. F_1 females are made pseudopregnant by breeding to vasectomized F_1 males, e.g. (BL/6 x BALB/c)F_1. (Prior to mating with foster mothers, vasectomized males are placed together with fertile females for at least two weeks to verify sterility.) Unlike the breeding for embryos by natural mating, the age of the females for foster mothers is perhaps more critical. This is primarily because older females are generally larger and appear to tolerate the surgery and pregnancy better than younger, smaller mice. However, the smaller mice (<24 g) can be easier to operate on (e.g. finding the ovaries) because of the amount of fat often found in bigger mice. Consequently, we use F_1 females older than 8 weeks of age and up to several months old weighing 22-30 g.

Foster mothers are needed as hosts for the injected/aggregated blastocysts/morulas and are used at day 2.5 days p.c. Using the example of beginning ES cell injections on Monday, F_1 females are placed together on Friday (again, two F_1 females/stud male) with plug check on Saturday. Plug-positive females are separated into a new cage and are used late Monday for transfers. The number of breeding cages required for generating foster mothers is less than that for embryo preparation and depends on the number of blastocysts to be transferred. Ideally ~12-16 blastocysts are transferred to a single foster mother which means 6-8 per uterine horn. It is possible to transfer fewer blastocysts although a minimum of six blastocysts (injected or not) should be transferred per uterine horn and the transfer of only one uterine horn is possible and sometimes even preferable. Usually, set up 10 - 20 cages (one vasectomized male, two F_1 females each) depending on whether you expect to need only one or several pseudopregnant females for transfer (judged by the number of plug positive females obtained for blastocyst production the same day). Often there is a considerable day-to-day variation in the number of plugged F_1 females. Whenever possible use females derived from different storage cages since the ovulation of females within one cage tends to be synchronized. However, plug-positive F_1 females that are not used for transfer can be used again for breeding after a two week period.

After blastocyst transfer foster mothers should be kept in a quiet, separated environment in the animal facility as stress (and also the smell of males or foreign strains) may reduce the number of successful pregnancies.

When checking plugs in mice bred for blastocysts vs mice bred for foster mothers you deal with uncertainties differently. That means for a questionable plug positive female (not 100% certain) then the rule of thumb would be to include her with the plug-positive females for blastocyst preparation but **DO NOT** include her with plug-positive females for foster mothers.

Vasectomy

For the production of pseudopregnant foster mothers you need a stock of at least 20 vasectomized males. F1 males, e.g. (C57BL/6 x BALB/c)F_1 or (C57BL/6 x CBA)F_1, or outbred males should be used as their breeding performance is much better compared to males of most inbred strains. The males are vasectomized at young age (6 - 8 weeks) and can be used up to the age of one year.

Vasectomy belongs to the methods which cannot be truly learned by reading and needs direct demonstration and supervision by an experienced person. What follows here is a brief description of the procedure which may help you to memorize the necessary steps. For introduction into the anatomy of the male reproductive system see Rugh (171).

1. Weigh, anaesthetize a male with avertin or ketamine/xylazine (Appendix A), place dorsally on the bench and wet abdomen with 70% ethanol. Make a transverse, ventral skin incision (~1 cm) about 1 cm rostral of the penis. Open the peritoneum with a similar incision.

2. Push the scrotum with your finger upwards into the direction of the incision. Lift the peritoneal wall with one pair of forceps, locate the testicular fat pad, grasp the fat pad with another forcep and pull out the testis. Locate the vas deferens without confusing it with the epididymis.

3. When identified, grasp the vas deferens with forceps and lift up such that a loop is formed. Take another pair of forceps, heat the tips in a gas flame until red and grasp the stem of the loop thereby interrupting the vas deferens and cauterizing the ends (see Figure 23). Separate the ends if they stick to each other. Return the testis to the peritoneum and repeat the procedure with the second testis.

4. Close the incision of the peritoneal wall with surgical silk suture and seal the skin incision with one or two wound clips (Appendix D). Let the vasectomized males recover for two weeks, remove wound clips and test for sterility by mating to normal females for 2 - 3 weeks.

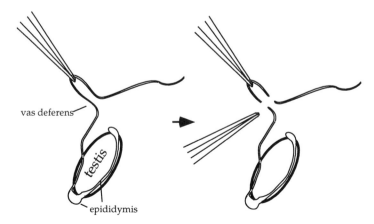

Figure 23. Cauterization of the vas deferens in the generation of sterile male studs.

Chapter 24
Injection, holding, and transfer needles

If you will inject ES cells into blastocysts, then before injections begin you will need to prepare your own stock of needles for embryo manipulation. You should devote as much time as necessary beforehand to generate a comfortable stock of all needles. As equipment for the production of microneedles you need a (simple) needle puller (e.g. Kopf 720), a stereomicroscope (dissecting scope; e.g. Wild M5A, 15X oculars), a microforge (e.g. de Fonbrune, Bachofer), a microscope with 400x magnification and a bunsen burner or ethanol burner.

The needles required are:

1. Transfer needles - for manipulation of isolated embryos. Transfer needles (Appendix C) are made by flaming a capillary in the center and, when the glass starts to melt, simultaneously pulling and twisting slightly. Those which have been pulled appropriately are broken at the pulled ends and the two resulting needles are examined in the dissecting scope for a relatively straight (not jagged) end and for appropriate diameter. A transfer needle of the right size can be placed alongside the new needle as comparison for appropriate diameter which should be ~1-1.5x the diameter of a blastocyst. The tips of these needles are then slightly flamed to remove the sharp edge. For the manipulation of blastocysts with the transfer needle, you must attach the needle to a mouth pipet with flexible tubing interrupted by a 0.22 µm sterile filter (Figure 24).

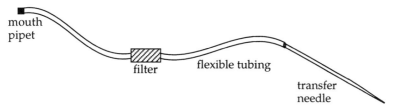

Figure 24. Mouth-controlled pipet for embryo handling and transfer.

The actual isolation, manipulation, injection, and transfer of blastocysts is carried out in a clean room isolated from traffic and drafts.

2. Injection needles - mounted on injection microscope and used for gathering ES cells and injecting into blastocysts. Injection needles (borosilicate glass capillaries 1.2 mm O.D. x 0.69 mm I.D.; standard wall without filament; Appendix D) are made by the use of a needle-puller which pulls the needles to an appropriate length (2-3 cm) and diameter (Figure 25).

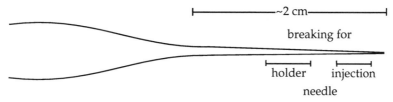

Figure 25. A drawn-out capillary pipet is shown together with the regions to break for generating injection and holding needles.

After pulling a number of needles they are next taken to the dissecting scope (~75x magnification) where a fresh scalpel blade is used to physically 'cut' the needle near the tip in the region of approximately 15 μm diameter. This is achieved by holding the needle firmly against the platform (e.g. covered by a piece of paper) and, while observing in the scope, slowly using a brushing motion, bringing the scalpel edge closer and closer to the needle until a break is made. The break is then examined for an appropriate tip and diameter. The edge of the needle should appear smooth with a sharp tip as shown in Figure 26. Needles which appear good are placed aside to be scrutinized later under higher magnification.

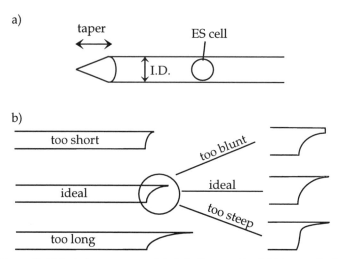

Figure 26. Criteria for an appropriate injection needle. (**a**) The needle should comfortably accommodate the diameter of an ES cell (which may vary with the ES cell line) without being too large as to damage the blastocyst upon injection. (**b**) Also note that the tip should have a taper that is relatively smooth and of appropriate length.

From a single needle several attempts can be made at creating a good tip until the diameter is too large for use as an injection needle. These needles can be set aside and used again to make holding needles (see below). Once a number of potentially good needles have been made these are examined under 400x magnification to assess the quality of the needle tip. Additionally, the scope ocular is equipped with a scale which is used to examine the diameter of your needle. The inside diameter (I.D.) at the tip should be approximately the same size as the ES cells which you will inject (~10-15 µm) and a taper which is slightly longer. Usually one out of 30 - 40 breaks will result in a good injection needle. In a more controlled method for the production of injection needles the tip is first beveled in a beveling device and then sharpened using a microforge (see ref. (16) for description).

Needles with too large a diameter will damage the blastocyst upon injection and with too small an I.D. diameter will be nice for injection of the blastocyst but will physically stress and damage the ES cells during collection and expulsion.

For collection of ES cells the injection needle must be inserted into the holder such that the opening is oriented towards the bottom of the injection chamber. When looking at the tip through the microscope, it is often difficult to determine whether the opening is facing up or down. Attempts to collect ES cells with an opening in the wrong orientation will (sooner or later) be evident. Rotate the needle 180° (gently mark the needle for orientation) and try again.

3. Holding needles - mounted on injection microscope and used for holding the blastocysts as you inject ES cells. Use injection needles which have been discarded (because the proper size diameter has been exceeded) or freshly pulled capillaries (same as injection needle capillaries). A new razor cut is made at a larger diameter with the aim being ~80-100 μm outer diameter. Either a nice straight break (rare) is required at the appropriate diameter or the irregular broken tip is melted at low temperature to a glass bead which has been attached to the tip of the microforge filament. When the tip of the needle sticks to the glass bead the filament is switched off and the needle will break with a straight tip by the contraction of the filament. Once achieved, this tip is polished and narrowed to ~10 μm by heating closely to the tip of the microforge filament (Figure 27).

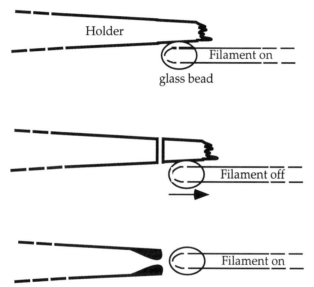

Figure 27. Use of a microforge in making blastocyst holding needles.

Chapter 25
Blastocyst injection

Injection apparatus

Generally, a microscope with up to 300-400x magnification, two micromanipulators, two micrometer syringes, tubings, needle holders and injection chambers are needed for blastocyst injections. For the collection and handling of blastocysts a stereomicroscope (~50x magnification) is needed and a 37°C incubator (preferably with CO_2 supply) for blastocyst culture. Since there are many suppliers of micromanipulation instruments many different configurations are possible. The set-up we use, and regard as comfortable, consists of a Leitz Labovert inverted microscope equipped with 10x oculars and 10x, 16x, 32x and 40x phase contrast objectives. This scope is mounted on a Leitz ground plate together with two Leitz micromanipulators with Leitz or Eppendorf needle holders. As syringes we use AGLA micrometer syringe holders (Vetter) equipped with Hamilton syringes (100 µl to control injection needle, #710 or #1710, Hamilton; 250 µl to control holder needle, #725 or 1725) which are connected via thick-walled tubing to the needle holders (micrometer syringes can be also purchased from Narishige). The system is filled with paraffin oil (Merck (Germany) # 7174) and interrupted by a three way valve connected to an oil filled reservoir syringe. We do not use a cooling stage for the injection chamber and work with straight needles inserted into the injection chamber at an angle of 10 - 25°.

Blastocyst preparation/injection

First thing in the morning remove EF-medium/HEPES, take out (MT)PBS, trypsin, and EF-medium to warm up as you retrieve the plug-positive females (for blastocysts) and check plugs (for blastocysts and foster mothers) for future injection. If beginning a week of injections you should prepare enough EF-medium + HEPES (33.3 mM, pH 7.2; Gibco) for the week. DNase (300 Kunitz Units/ml; Sigma, bovine pancreas type IV) is added to this EF-medium + HEPES to be used in the injection chamber.

This media can be prepared in a large amount and stored at -20°C. Aliquots of 10 ml DMEM-EF/HEPES and 2 ml DMEM-EF/HEPES/DNase are convenient amounts to use. The 10 ml aliquots will be used for flushing, transferring, and holding blastocysts while the 2 ml aliquots will be used for setting up injection chambers.

1. Plug positive mice taken at day 3.5.

2. Soak abdomen in disinfectant (70% ethanol) to sterilize and cut back skin/peritoneal membrane. Lay aside intestines and locate uterine horns. Grab the fat on the ovary and lifting slightly cut away fat along the fallopian tubes. Cut just below the fork and below ovary and wash horns in dish of (MT)PBS. Remove remaining fat and cut away second ovary (transferring to a new dish of (MT)PBS.

3. Withdraw 1 ml EF medium containing HEPES buffer (150 ml EF media + 5 ml 1 M HEPES, Gibco) into 1 ml syringe with 26G needle. Holding the end of the horn with forceps, insert needle into tube, grasping slightly with forceps, and flush with 0.5 ml media into a 9 cm dish (Figure 28). Repeat with other horn taking care to leave horns submerged in media as to not squirt out media (and blastocysts). Surveying the dish for blastocysts may be facilitated by marking the bottom of the dish into sections.

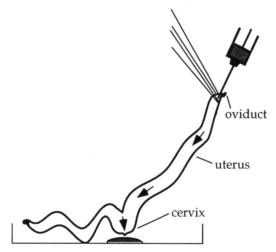

oviduct

uterus

cervix

Figure 28. Flushing preimplantation embryos from the uterine horn into a dish.

4. Set up two drop cultures with ~100 µl EF/HEPES buffer/drop in 35 mm dishes. Overlay with paraffin oil for embryo cultures (Sigma, embryo tested, #8410 or Merck #7161). One drop is used for good blastocysts which will be kept at room temperature and the other is for morulas which will be left in the incubator at 37°C (and preferably

107

with 7.5% CO_2, if available) and may develop into usable blastocysts with time.

5. With mouth pipet (and filter) use transfer needle to find and sort blastocysts. If the manipulation of blastocysts is initially difficult using the mouth pipet, then a 'buffer' can be created which decreases the sensitivity of movement. This is accomplished by taking up into the transfer needle ~1 cm of mineral oil => a small air bubble => ~1 cm of media => a small air bubble => media (Figure 29).

Figure 29. Buffering movement in a transfer needle with air and medium.

Under dissecting scope find and remove all blastocysts and morulas transferring them to the appropriate drop culture.

7. Trypsinize ES cells and resuspend in EF medium without 2-β-mercaptoethanol (e.g. from single well of 12 well plate resuspend in 2 ml) and take to injection room.

8. Prepare injection chamber and place ~200 µl injection medium (EF/HEPES/DNase) in the center. As an injection chamber we simply use a plexiglass slide (55 x 40 x 2 mm) with a hole of 20 mm diameter in the center. A 24 mm glass coverslip is then placed over the hole and secured with tape creating an injection chamber of 2 mm depth. Spread the medium outwards leaving room on the perimeter. The height of the injection medium should not be above the injection chamber. Overlay with embryo paraffin oil using a syringe with needle.

9. Using a transfer pipet, withdraw ES cells and place in injection chamber and then add blastocysts. The organization of the injection chamber depends on personal preference but the goal is to have an area for ES cells, an area where the blastocysts are located, and an area where injections are performed and injected blastocysts are stored. For example, distributing the ES cells in a circle around the perimeter or along one side of the injection chamber and placing blastocysts in center (Figure 30). Place the injection chamber on the injection scope stage, focus on the chamber bottom at low magnification, immerse injection and holder needles (oil filled) until located on the ground of the plate in injection medium. For starting injection aspirate medium in both needles such that you can always see the medium/oil interface.

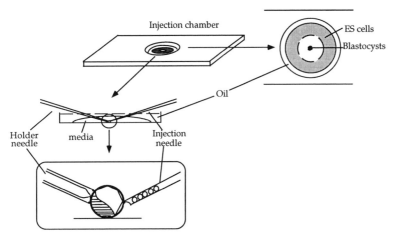

Figure 30. Organization of the injection chamber with respect to ES cells, blastocysts, and uninjected blastocysts.

10. With B6III ES cells it is recommended to place ES cells at 4°C while injecting (for ~1 hour or so). This greatly improves the quality of the ES cells for injection. Nevertheless, fresh ES cells should be trypsinized if injection continues for more than a couple of hours.

11. Learning how to inject blastocysts by reading is a long way from reality and is, therefore, not included in this primer. For this there is really no substitution for actually doing it although watching an experienced person injecting on a monitor is certainly recommended.

   Nevertheless, what follows is a very brief and far-removed attempt to describe blastocyst injection. ES cells are first gathered (at ~200-300x magnification) with the injection needle by taking up individual small, refractile, healthy-looking ES cells and avoiding any debris. This is shown in Figure 31 and accomplished by first aligning the opening of the injection needle above an ES cell, and then lowering the needle at the same time as providing suction to take up the ES cell (taking care to have the needle with the opening in the right orientation).

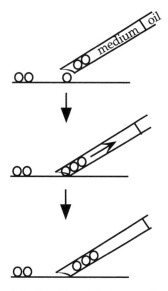

Figure 31. Collecting ES cells with an injection needle.

Either 10 - 15 cells for one, or 20 - 30 cells for two injections (or more) are collected such that the cells are located in a row close to each other at the tip of the needle. They must be separated from the mineral oil/medium interface by some medium free of cells as ES cells will lyse when in contact with oil. If this happens cellular debris will coat the inside wall of the injection needle and this will make things significantly more unpleasant because the control of the needle will become difficult due to the oil sticking to the debris. Dirty needles can be occasionally cleaned and rescued by pushing out oil at high pressure for a few minutes and/or by *very carefully* aspirating and ejecting 70% ethanol several times making sure not to create any oil/ethanol bubbles as this will just compound your problem. If debris is located outside try to remove using the holder needle or wiping gently with a tissue soaked in 70% ethanol. Once enough ES cells are within the needle then a blastocyst is taken up by the holding needle and moved to an area of the chamber where injections will take place. (This should be some place where injected blastocysts will not be confused with uninjected blastocysts.) The blastocyst is then manipulated such that a cell border between the trophoblast cells (node) is directly opposite the point at which the blastocyst is firmly held by the holding needle (Figure 32). For injection the blastocyst must lay on the bottom of the injection chamber and may even be pressed against the surface slightly since it easily slides away when pressure from the injection needle is applied. Now the injection needle, loaded with ES cells, is brought to the blastocyst and aligned with the node (see Figure 32). This is often a tricky part of injecting since you must align, in three dimensions, the very tip of the injection needle with the thinnest area of the blastocyst (at the node). Finally, with a controlled forcible movement, the injection needle is inserted into the blastocyst (at 300-400x magnification) and the ES cells are released. Care should be taken not to drive the injection needle too far into the blastocyst as this can result in the breaking of the injection needle against the holding needle. In general, ~15 ES cells are injected per blastocyst. Because ES cells die in the injection chamber within ~2 hours of collection, the collection of ES cells become more and more difficult with time. You need then to prepare a new injection chamber either with freshly trypsinized ES cells or cells kept at 4°C during the first session of injection. After injection blastocysts should be transferred soon (e.g. in groups of 10) to the incubator until transferred to foster mothers.

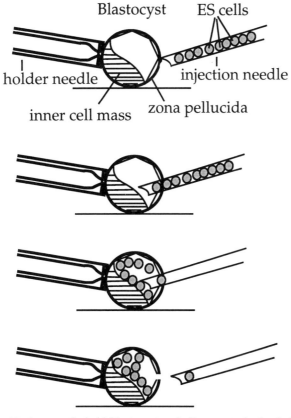

Figure 32. A correctly held blastocyst and alignment of a loaded injection needle at the zona pellucida. Note that the injection needle will pass between two trophoblast cells directly opposite the holding needle.

111

It is important to learn what injecting should be like under the right conditions. The 'right conditions' means having good needles, both injection and holding, and an injection scope which is functioning properly. Much time is often wasted simply fighting an improperly tuned machine or using bad needles. The time taken to remedy the machine is quickly regained in more efficient injections.

Chapter 26
Embryo aggregation

For embryo (morula) aggregation less equipment is needed than for blastocyst injection: required are a stereomicroscope (~40x magnification) with transmission light for setting up aggregates and a CO_2 incubator for culture so that researchers setting up a new lab may choose aggregation instead of blastocyst injection for the production of chimeric mice. For embryo aggregation eight-cell stage embryos up to morulas should be isolated from the oviduct of superovulated or normally mated females at day 2.5 p.c. After removal of the zona pellucida the embryos are placed individually, together with a small clump of about 10 - 20 ES cells, into small 'aggregation wells' of a plastic dish for aggregation (aggregation plate). After overnight culture the majority of these aggregates develop into compacted morulas or early blastocysts which are then transferred into the uteri of pseudopregnant females in the same way as microinjected blastocysts (see Chapter 27). Our protocol is adapted from the one described by Nagy and Rossant (170) except that only one embryo is used per aggregate and that media contain a low concentration of phenol red.

Using this protocol, germline chimeras can be obtained from aggregates of E14.1 ES cells with eight-cell stage embryos and morulas from C57BL/6 or CB.20 mice but also in combination with embryos from the CD1 outbred strain. Apart from the less expensive instrumentation, the main advantage of embryo aggregation is that it can be easily scaled up since an experienced individual is able to set up several hundred aggregates per day (compared to the injection of ~70 blastocysts/day). If embryos from inbred strains are used this difference is less important than it seems since the number of embryos recovered from 10 - 20 females (3 - 4/ female) is often limiting and can be also handled by blastocyst injection. Thus, for the large scale production of chimeras we recommend the use of the CD1 outbred strain since large numbers of embryos can be obtained at low expense from CD1 mice (~14/superovulated female). The frequency of germline chimeras obtained by aggregation of E14.1 ES cells and CD1 embryos is in the range of 40 - 60% (K. Pfeffer, personal communication) and comparable to the efficiency of blastocyst injection.

Starting an aggregation experiment requires the following preparations:

Three days before aggregation the mating of mice for the production of embryos must be started. Plug positive females are sorted out the next day (day 0.5 after plug) and used at day 2.5 p.c. When performing superovulation PMS must be injected five days before aggregation and hCG at the day of the set up of the breedings (see Chapter 23).

Two days before aggregation the ES cells, passaged at least once after thawing, must be plated into dishes without embryonic feeder cells and females must be housed together with vasectomized males for the production of foster mothers. Other required materials are: modified M2 and M16 media (Appendix A), Tyrode's (or Pronase) solution (Appendix A), embryo and ES cell transfer needles (Chapter 24), flushing and darning needles (see below), tissue culture dishes and anesthetics (Appendix A).

Isolation of embryos

The females serving as embryo donors must be processed in the morning of the second day after plug (day 2.5 p.c.) at the time when most of the embryos are at the eight-cell stage and located in the oviduct and upper uterus. The embryos are recovered by flushing the oviduct with M2 medium through its proximal opening (infundibulum) and in the next step the surrounding zona pellucida is removed. When embryos are manipulated outside of a CO_2 incubator as for isolation, setting up aggregation and embryo transfer they should always be kept in M2 medium, which is buffered by HEPES. M16 medium, buffered by CO_2/Carbonate, is used only for overnight culture in a 7.5% CO_2/air atmosphere as this medium rapidly becomes basic (red color) at the bench. See Appendix A for the preparation of modified M2 and M16 which should be used for embryo aggregation. For all procedures tissue culture plastic dishes are used (e.g. 10 cm (diameter) plates from Falcon, No. 3003). For covering cultures 'embryo tested' oil (Sigma, M8410) should be used.

1. Sacrifice females by cervical dislocation, soak abdomen in disinfectant (70% ethanol) to sterilize and open abdomen by incisions on the left and right side. Under good illumination (preferably from a cold light lamp), lay aside intestines and locate the ovary. Lift the ovary by grasping the ovary associated fat pad with Dumont No.5 forceps in one hand, cut off associated membranes and fat and then cut between ovary and oviduct (1. cut, Figure 33A) using a fine scissor with the other hand. Lift the remaining oviduct/uterus by grasping the proximal uterus, cut the uterus ~3 mm behind the entrance of the oviduct (2. cut, Figure 33A) and collect the isolated tissue in a large drop of M2 medium.

2. When all oviducts are collected, start to isolate the embryos by processing groups of ~10 oviducts. For each oviduct place a 20 μl drop of M2 medium on the lid of a 6 cm dish and transfer one oviduct into each drop by grasping the attached uterus piece with forceps. Do not touch the oviduct directly with forceps as it is easily destroyed. For flushing of the oviducts a 1 or 2 ml syringe filled with M2 medium and a blunt ended fine metal (flushing) needle is required. Suitable needles can be purchased from Hamilton (No. 90033) or self-made from 32 gauge injection needles by clipping off the sharp end and polishing of the tip. Place the lid on the stage of a stereomicroscope and locate (under transmission light) the opening of the oviduct (infundibulum), hold it in position with Dumont No.5 forceps and insert the flushing needle with the other hand until the tip reaches the first turn of the oviduct coil (be careful not to penetrate the oviduct wall; Figure 33B). Fix the needle inside of the oviduct by gently embracing with the forceps a part of the oviduct in which the needle is inserted (otherwise it will fall off the needle during flushing; Figure 33B). Flush the oviduct with ~100 μl of M2 medium. The oviduct should swell briefly during washing. If the flushing needle penetrated the wall the oviduct cannot be flushed any more but you may try to recover the embryos by the complete disruption of the oviduct using a pair of forceps. Remove the flushed oviduct from the M2 drop and proceed until all oviducts on the plate are processed. Flush only the number of oviducts which can be processed in 10 - 15 minutes as the small medium drops dry out rapidly and set up a new plate for the next group of oviducts.

3. For the collection and washing of the embryos place three 200 μl drops of M2 medium in a 6 or 10 cm dish. Pick up the embryos from the medium droplets using a mouth controlled embryo transfer needle and collect them in the first drop of M2 medium until all oviducts are processed. For transfer into the second drop of M2 select under high magnification all eight-cell embryos and morulas (Figure 33C) while leaving behind the debris and other developmental stages. Repeat this procedure once and collect the embryos in the third drop of M2 medium. Set up a 6 cm plate with a single 100 μl drop of M16 medium covered by oil. Transfer the embryos from the M2 medium into the M16/oil culture and keep the plate in the incubator (37°C, 7.5% CO_2) until you start to remove the zona pellucida. Before zona removal set up your aggregation plate.

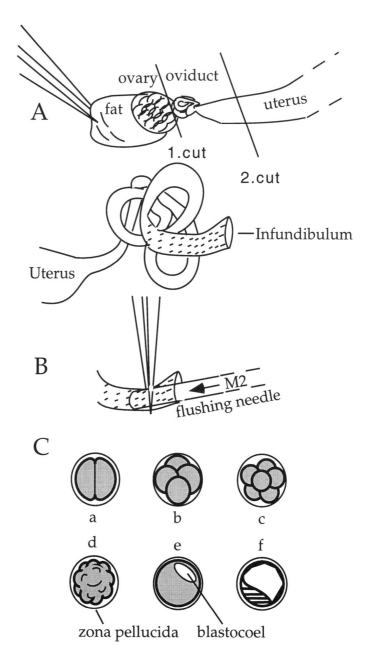

Figure 33. Isolation and appearance of embryos for embryo aggregation. **(A)** Upper part of the female reproductive tract (ovary, oviduct, uterus) showing the position of cuts made for the isolation of the oviduct. The lower part shows the isolated

oviduct with its opening (infundibulum) as used for embryo isolation. **(B)** Insertion of a flushing needle into the infundibulum. **(C)** Mouse preimplantation embryos at different stages of development: (a) 2-cell stage, (b) 4-cell stage, (c) 8-cell stage, (d) compacted 16-cell stage/morula, (e) early blastocyst, (f) expanded blastocyst.

Aggregation plate

This plate is used for the assembly and overnight culture of aggregates between ES cells and morulas. Using a darning needle, small crater-like depressions (aggregation wells) are indented into the plastic surface of the dish, each holding one embryo and a clump of ES cells.

1. Place rows of drops of ~20 μl M16 medium in a 10 cm dish. In the example shown in Figure 34 three rows of five (aggregation) drops are placed between an upper and a lower row of three drops. The latter drops serve to hold selected ES cell clumps before their final transfer into the drops for aggregation. Cover the plate with oil and generate six depressions (aggregation wells) in each aggregation drop by pressing a darning needle (sterilized with 70% ethanol) into the surface of the plastic dish. Each well is then suitable for the aggregation of six embryos and the plate shown in Figure 34 can be used for 90 embryos in total. Specially designed aggregation needles are available from BLS (see Appendix D), otherwise a number of conventional blunt ended darning needles should be purchased and tested for performance. While the bottom of the aggregation wells should be flat and approximately the diameter of the embryos to promote their contact with the ES cells the walls of the wells should not be too steep and deep (Figure 36) to facilitate the recovery of the embryos on the next day.

2. After completion of the aggregation plate keep it in an incubator (37°C, 7.5% CO_2) for equilibration until the zona pellucida is removed from the embryos.

117

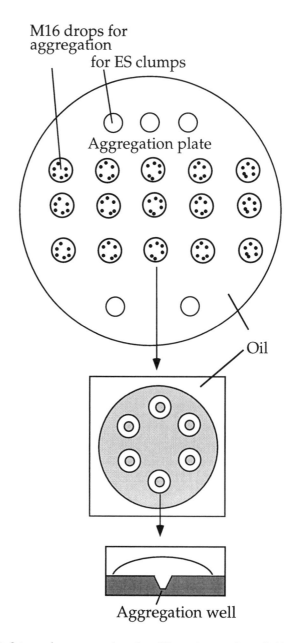

M16 drops for aggregation
for ES clumps

Aggregation plate

Oil

Aggregation well

Figure 34. Set up of an aggregation plate (10 cm tissue culture dish) for embryo aggregation. Drops of M16 medium (20 µl) are placed into rows and covered by oil. Using a darning needle for each m16 aggregation drop, six indentations (aggregation wells) are made into the surface of the dish. Below: top and side view of aggregation wells.

Removal of the zona pellucida

The zona pellucida can be removed either at low pH in acidic Tyrode's solution or enzymatically using protease (pronase). The chemical method is faster and more commonly used than the enzymatic digestion but requires more exact timing to avoid damage to the embryos.

Tyrode's solution

For a group of 10 - 30 embryos, place on the lid of a 10 cm dish a vertical row of 20 - 50 µl drops of M2 medium (one drop), three drops of Tyrode's solution (Appendix A), two drops of M2 and one drop of M16 medium (Figure 35). These drops need not to be covered with oil if the embryos are processed immediately after setting up the plate. Collect the embryos from the storage plate out of the M16 medium and place them in groups of 10 - 30 into the M2 drops at the top of the rows of the plate with Tyrode's solution. Then process only one group of embryos at one time as the timing is critical. Take up the first group of embryos in a minimal volume of medium using a mouth controlled embryo transfer needle and wash through the first two drops of Tyrode's solution. Washing should involve the transfer of the embryos into the new drop and subsequently the complete mixing of the media from the pipette and the new drop. Transfer the embryos into the third drop of Tyrode's and observe until the zona is completely dissolved. This process should be completed within two minutes after transfer into Tyrode's solution. Immediately after zona removal wash the embryos through the lower two drops of M2 and once in the drop of M16 medium. After zona removal the embryos become very sticky and it is best to space them at a certain distance in the embryo transfer needle and to distribute them directly into the wells of the aggregation plate (one embryo per well). When all embryos are processed keep the aggregation plate in the incubator (37°C, 7.5% CO_2) until the ES cells are prepared for aggregation.

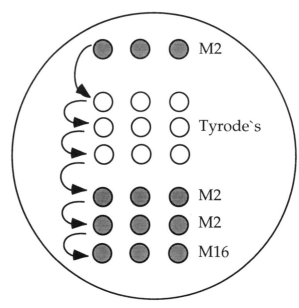

Figure 35. Set up of a 10 cm tissue culture dish for the removal of the zona pellucida in acidic Tyrode's solution. See text for further explanations.

Pronase method

Using a mouth controlled transfer needle collect all embryos from the M16 storage plate, transfer into a large drop (e.g. 200 μl) of M2 containing 0.5% pronase (see Appendix A) and mix by pipetting. Incubate the embryos for about 5 minutes or more at 37°C until the zona has disappeared. Wash the embryos through two 200 μl drops of M2 and one drop of M16 medium. Finally distribute the embryos into the wells of the aggregation plate by placing one embryo per well. Keep the aggregation plate in the incubator (37°C, 7.5% CO_2) until the ES cells are prepared for aggregation.

Preparation of ES cells

ES cells cultured under standard conditions should be plated two days before aggregation on 10 cm (non-gelatinized) tissue culture dishes (e.g. Falcon, No. 3003) without embryonic fibroblasts to reduce the number of fibroblasts during aggregation. ES clones used for aggregation should be continuously propagated on embryonic fibroblasts but plates free of fibroblasts should be prepared in addition for each day of aggregation. After thawing ES clones should be passaged at least once on embryonic fibroblast before plating for aggregation. At the day of aggregation, when the embryos are distributed into the wells of the aggregation plate, wash the ES cell plate once with PBS and add 2 ml of Trypsin solution (or modified Trypsin solution, depending on the ES line). Keep the plate at room temperature and observe the trypsinization process in a microscope.

When the colonies start to detach from the plate (test by gentle rocking of the plate) add 5 ml of EF medium (Appendix A) and gently pipette up and down to break up larger aggregates. The goal is to limit the trypsinization such that a part of the ES cells is still loosely connected in clumps of about 10 - 20 cells.

Embryo aggregation

As soon as the ES cells are trypsinized the actual aggregation can be set up using an ES cell transfer needle for handling the ES cell clumps. These capillaries are the same as used for handling embryos (Chapter 24), except that the diameter of the tip should be smaller (about half of the diameter of an embryo) to facilitate the manipulation of the small ES cell clumps. Place the plate with trypsinized ES cells under a stereomicroscope at ~ 40x magnification, pick up a number of clumps consisting of 10 - 20 ES cells and transfer these clumps into the outer drops of M16 on the aggregation plate. Avoid clumps which also contain fibroblasts (large cells). When a sufficient number of clumps (e.g. twice the number of embryos) is transferred into these M16 drops, make the final selection of ES clumps for aggregation under the highest magnification. Transfer six suitable clumps into the middle of each of the M16 drops containing six embryos in the aggregation wells. Using a mouth controlled ES cell transfer needle, take up an individual ES clump and gently blow it out of the needle above one of the embryos so that it will lie on or in close contact to the embryo after settling down in the aggregation well. Alternatively the ES clump may be simply pushed over the wall of the aggregation well with the tip of the pipette so that it comes into close contact with the embryo after sliding down the wall. In any case more than one cell of the ES clump should be in contact with the embryo. When all aggregates are set up return the aggregation plate to the incubator (37°C, 7.5% CO_2) and culture overnight. The next day 70% or more of the aggregates should have developed into morulas or early blastocysts and integrated the ES cells. These aggregates are now ready for transfer into pseudopregnant females at day 2.5 p.c. (see Chapter 27). If the ES cell clump is still visible the aggregation was unsuccessful and embryos from such wells should be discarded. Collect the successful aggregates immediately before embryo transfer (in the morning or afternoon) and place them for transportation into a 100 µl drop of M2 in a plate covered by oil.

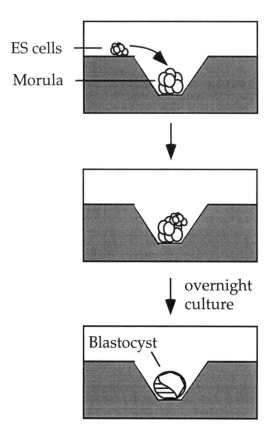

Figure 36. Aggregation of an 8-cell stage embryo with ES cells in an aggregation well (side view). Using a transfer needle (not shown), a clump of 10 - 20 ES cells is placed on (or near) an embryo. During overnight culture the ES cells develop together with the embryo into a blastocyst (or morula, not shown).

Chapter 27
Blastocyst transfer

After a days work of injecting blastocysts you still must transfer these embryos to foster mothers. Initially, this can be a source of anxiety but with time can also become quite enjoyable.

1. For blastocyst transfer you will need:

   • Surgical instruments (2 fine forceps (e.g. Dumont No.5), 1 fine and 1 large scissors, a serafin clip, a sewing or hypodermic needle with holder, 9 mm wound clips and clip applicator (e.g. Clay Adams, Becton Dickinson), surgical suture (e.g. Ethicon), needle and holder.

   • transfer needles (≥2 needles)

   • mouth pipette/tubing

   • 1x Avertin (or other anesthetic)

   • injected blastocysts in drop culture

2. Set up and clean surgical equipment.

3. Prepare cage(s) and weigh foster mother(s).

4. Load transfer needles with blastocysts. Place two ~200 µl drops of EF/HEPES or M2 medium in a 60 mm dish without covering by oil. Transfer blastocysts used for one transfer from drop/oil culture to one of the freshly prepared medium drops. Take a new transfer needle (free of oil outside) and fill the tip by mouth control with oil until movement is smooth. Remove oil sticking outside at the tip of the needle with tissue paper. Take up a small air bubble (1 - 2 mm), medium (3 - 5 mm) from the drop containing no blastocysts, another air bubble followed by ~1 mm medium. Now take up ~6 blastocysts from the other drop such that the embryos are located in ~1 mm medium in a row at the very end of the needle (Figure 37). Usually transfer at least 6 blastocysts per uterine horn, therefore if enough blastocysts have not been injected then make up the difference with

uninjected blastocysts. Maximum should be 9-10 blastocysts per side (18-20 blastocysts per mouse). Take extra transfer needle(s) with you to the animal facility in case of accidental breakage.

5. Inject foster mother with Avertin or xylazin/ketamin i.p. (Appendix A). If anesthetized properly then mice should fall asleep within ~1 minute. If this does not occur then either the anesthetics may have lost some potency or may not have injected intraperitoneally (e.g. injected into fat tissue). At this point it is perhaps best to retrieve another foster mother but if this is not an option then you may try injecting more anesthetics. How much more is difficult to guess although 10-15% more would be reasonable. The concern is that it is relatively easy to administer a lethal amount and/or the anesthesia may interfere with the pregnancy.

6. The transfer surgery, like blastocyst injection, must be demonstrated to you by an experienced person and is learned by doing, not by reading. It will be described only briefly here. For anatomy of the female reproductive tract see Rugh (171). Obviously, observing the transfer is an important initial step but even this is not the same as actually performing the task yourself. In the beginning virtually everyone experiences a certain amount of anxiety which is natural. This is exaggerated when beginning injections simply because the transfer surgery comes at the end of a very long day. At this point physical fatigue would be expected although the opposite occurs - an adrenalin rush. During the surgery it is important to retain some perspective during the transfer, i.e., life will go on after a poor transfer. The words presented here are merely symbolic as it is much harder to accomplish in practice, especially after a full day of injecting ES cells and the blastocysts by this time have become incredibly precious. Nevertheless, with time the transfer can be the most pleasurable part of the day. Leaving the lab after a day of injecting and a final transfer into the host (there is not a better place for those blastocysts to be) can be an extremely satisfying feeling.

Blastocyst transfer

•Place anesthetized foster mother, lying ventrally on the lid of a 90 mm dish, on the stage of the stereomicroscope with her head oriented away from you. Illuminated by a (two armed) cold light lamp, wipe the back with 70% ethanol and part the fur (~1 cm) above the spine ending about at the height of the hind knees. Now make a 5 - 10 mm incision through the exposed skin cutting as little hair as possible.

•Using fine forceps slide the incision to the left until the pink-reddish ovary is seen just beneath the peritoneum. Make a ~5 mm incision in the peritoneum avoiding blood vessels and, grasping the fat pad of the ovary, pull out the ovary, oviduct and upper part of the uterus. Clamp the fat pad with serafin clip (being careful to avoid the ovary) to fix the position of the uterus.

• Embrace, with minimal pressure, the uterus with a fine forcep and penetrate the uterus wall with the sewing needle using your other hand (Figure 37). Now holding the uterus epithelium with the forcep just in front of the hole in the uterus wall, lay the sewing needle holder aside, take up the transfer needle and insert the tip 3 - 5 mm into the uterine lumen. Blow out the medium containing the embryos (not more) by mouth control observing the air bubble in the transfer needle outside of the uterus. If the needle is blocked move back and forth as this generally means the opening is blocked by tissue. After ejecting the blastocysts, withdraw the needle from uterus. Remove the serafin clip and reposition the uterus, oviduct, ovary and fat pad back into the peritoneal cavity. Suture the incision of the peritoneum (optional if incision is small).

Transfer pipette

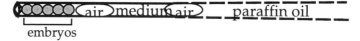

Uterus Transfer

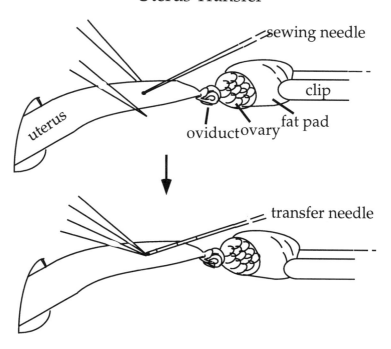

Figure 37. Loading embryos into the buffered transfer pipette and transfer into the uterus.

•Place the mouse aside and take up the blastocysts for the other uterus horn with the transfer needle. Put the mouse back on the microscope stage now head oriented towards you. Repeat the transfer procedure as described above. If not blocked with blood transfer needles may be used for several transfers in one day. Use, however, a new needle for each transfer day. Finally seal the skin incision with 1 or 2 wound clips taking care not to include the peritoneal wall or any hair.

•After transfer the animals should be kept warm to minimize heat loss. We simply cover them (except the head) with cage bedding.

Appendices

APPENDIX A

Reagents and Recipes

ES cell culture

EF medium:

> DMEM (high glucose without sodium pyruvate; GIBCO #41965-096)
>
> 60 ml FCS (10%); any tissue culture FCS is usable.
>
> 6 ml sodium pyruvate (Gibco, 100X, 11360-039, store at 4°C);
>
> 6 ml penicillin/streptomycin (Gibco, 100x, #15140-114; aliquot and store at -20°C).
>
> 6 ml L-glutamine (Gibco, 100x, #25030-024; aliquot and store at -20°C)

ES medium:

> DMEM (high glucose without sodium pyruvate)
>
> 75 ml FCS (15%); the batch is crucial and must have been tested for plating efficiency and toxicity of ES cells (see Chapter 14, Section 2).
>
> 6 ml sodium pyruvate (Gibco; 100x, 11360-039) store at 4°C); 1 mM final concentration.
>
> 6 ml penicillin/streptomycin (Gibco, 100x, #15140-114, store 6 ml aliquots at -20°C); final concentration 100 IU/ml penicillin, 100 µg/ml streptomycin.
>
> 6 ml L-glutamine (Gibco, 100x, 25030-024, store 6 ml aliquots at -20°C); final concentration 2 mM.
>
> 6 ml non-essential amino acids (Gibco, 100x, #11140-035; store at 4°C)
>
> 0.6 ml LIF (supernatant from LIF-transfected CHO cells; store aliquots of ~0.6 ml at -20°C)
>
> 0.6 ml 0.1 M 2-β-mercaptoethanol. Prepare by solving 50 µl 2–β-mercaptoethanol (Merck #15433) in 7 ml PBS, filtrate sterile and store aliquots of ~0.6 ml at -20°C. Alternatively purchase 50 mM 2-ME from Gibco (#32350-010) and use 1.2 ml per bottle of medium.

Gancyclovir; GANC sodium salt (Cymeven[R] from Syntex No. 115561)

Gelatin is from porcine skin / 300 bloom (Sigma no. G-1890; Type A); use at final concentration: 0.1% in H_2O or (MT)PBS (autoclaved).

LIF CHO transfected cells (8/24 720 LIFD(.1)) from Genetics Institute Cambridge Massachusetts; grow in α-MEM (without nucleosides) and methotrexate as recommended until plates are confluent. Remove methotrexate by washing 5x. Add fresh medium and place in incubator for 2 hours. Wash again 5x. Finally add fresh medium and harvest supernatant after 2 days.

Mitomycin C (Sigma #M-0503 from Strep. caespitosus = 2 mg); stock solution at 1 mg/ml; add 2 ml sterile (MT)PBS and aliquot in ~500 µl aliquots at -20°C. Use at 1:100 dilution (40 µl in 4 ml DMEM).

MT-PBS (mouse tonicity PBS) pH 7.3

4 mM $NaH_2PO_4 \cdot H_2O$	0.552 g
16 mM $Na_2HPO_4 \cdot 2H_2O$	2.848 g
150 mM NaCl	<u>8.766 g</u>
	H_2O to 1 liter

PBS

1.47 mM KH_2PO_4	0.2 g
8.1 mM $Na_2HPO_4 \cdot 2H_2O$	1.44 g
2.7 mM KCl	0.2 g
137 mM NaCl	<u>8.0 g</u>
	H_2O to 1 liter

Transfection buffer:

HEPES	20 mM, pH 7.0
NaCl	137 mM
KCl	5 mM
Na_2HPO_4	0.7 mM
glucose	6 mM
2-β-ME	0.1 mM

to prepare transfection buffer mix:

stock solutions		volume
1 M NaCl	2.92 g/ 50 ml	6.85 ml
1 M KCl	3.73 g/ 50 ml	0.25 ml

0.5 M Na$_2$HPO$_4$ (anhydrous)	3.55 g/ 50 ml	70 µl
0.5 M glucose (monohydrate)	4.95 g/ 50 ml	0.6 ml
0.1 M β-ME	50 µl/ 7 ml PBS	50 µl
1 M HEPES (Gibco)		1.0 ml
pH to 7.0, add glass distilled H$_2$O to:		50 ml
		sterile filtrate

10x Trypsin/EDTA (in 1x saline; Gibco #043-054004 100 ml); dilute 1:10 to 1x trypsin (0.05%) with MT-PBS.

Modified trypsin/EDTA (0.05 % w/v) for BL/6-III and BALB/c-I ES cell lines (modified from (14, 151)):

> 0.5 g trypsin powder (Sigma, T-1005)
>
> 0.4 g EDTA
>
> 7.0 g NaCl
>
> 0.3 g Na$_2$HPO$_4$·12H$_2$O
>
> 0.24 g KH$_2$PO$_4$
>
> 0.37 g KCl
>
> 1.0 g D-glucose
>
> 3.0 g Tris
>
> 10 mg Phenol red (Sigma, P-3532)
>
> to 1 liter distilled H$_2$O; adjust to pH 7.6

sterile filter, aliquot and keep at -20°C. For daily use keep one aliquot at 4°C and add 1% chicken serum (Gibco, 16110-033) to provide a protein containing solution to the ES cells but which will not inhibit trypsin (chicken serum contains no trypsin inhibitor).

Embryo aggregation and transfer

Anesthetics

Avertin is commonly used for anesthesia in mouse surgery and is made up as a 40x stock solution. The stock solution is prepared by dissolving 10 g (5 g) of 2,2,2-Tribromoethylalcohol (99%, Aldrich No. T4840, 25 g) in 10 ml (5 ml) of tertiary amylalcohol (2-methyl-2-butanol; 99%, Aldrich No. 24,048-6, 100 ml) and stored at - 20°C protected from light. To prepare 10 ml of the 1x working solution add 250 µl of the 40x stock into a 15 ml tube,

129

add 9.75 ml PBS (warmed up in a water bath to ~60 - 70°C) quickly and mix forcefully by pipetting up and down. The working solution must be stored light protected (e.g. wrapped in foil) at 4°C and can be used for up to two weeks.

Avertin is applied to mice by intraperitoneal injection and the dose needed for 30 - 60 minutes of anesthesia is generally around (15 -) 17 µl/ gram body weight. Increase this dose by about 10% when using older, fat mice (> 30 g body weight).

An alternative to Avertin, used in veterinary medicine, is anesthesia by the combined application of ketamine (100 mg/ kg) and xylazine (10 mg/ kg). To prepare the working solution (10 mg/ml ketamine; 1 mg/ml xylazine) mix 1 ml of Ketanest^R 50 (50 mg ketamine/ml from Parke-Davis) with 4 ml of sterile saline (0.9% NaCl) and 0.25 ml of Rompun^R (2% xylazine solution from Bayer AG). Store at 4°C up to 2 weeks.

For anesthesia of mice inject 10 µl of the working solution per gram body weight intraperitoneal or use 250 µl for all mice with a body weight of 20 - 30 g.

For the use of other anesthetics and a general overview see the article by Flecknell (174).

Hormones for superovulation

PMSG and hCG hormone preparations are dissolved in sterile saline (0.9% NaCl) at 50 IU/ml and stored in aliquots for single use at -20°C for up to one year (175). For superovulation inject intraperitoneally 100 µl (5 U) per female. We purchase either PMSG (Intergonan^R) and hCG (Ekluton^R) from Vemie Veterinär Chemie GmBH (Kempen, Germany) or from Sigma (PMSG, No.G-4877; hCG, No. C-1063).

Modified M2 and M16 media

Since the quality of these media is critical for the development of embryo aggregates chemicals and water of high purity must be used. We use double distilled water suitable for cell culture and chemicals from Sigma tested for embryo culture. If you are not sure about the quality of your water you may compare it to tested water from a company (e.g. Sigma, W1503). For media preparation and storage cell culture plastic ware should be used. The substances are weighed into 50 ml tubes, filled up with water and filtered through 0.22 µm filters. To avoid contamination of the solutions with toxic substances from sterile filters the first 10 ml of filtrates should be discarded.

The recipes of the stock solutions are the same as in Hogan et al. (113) except for the ten-fold reduced concentration of phenol red, which has a weak estrogenic activity, in stocks B and E. Ordering numbers for chemicals refer to catalogues from 1996.

Stock solutions

Component	per 50 ml	Source	Storage
Stock A (10x)			
NaCl	2.767 g	Sigma, S7653	
KCl	178 mg	Sigma, P9933	
KH_2PO_4	81 mg	Sigma, P0662	
$MgSO_4 \times 7\, H_2O$	146.5 mg	Sigma, M9397	6 months
Sodium lactate	1.659 ml	Sigma, L7900	at - 80°C
(60% Syrup)			
Glucose	500 mg	Sigma, G6152	
Penicillin G	30 mg	Sigma, P4687	
Streptomycin	25 mg	Sigma, S1277	
Stock B (10x)			
$NaHCO_3$	1.0505g	Sigma, S6014	prepare
Phenol red	2 ml	Sigma, P3532	fresh
(50 mg/ml in water)			
Stock C (100x)			prepare
Sodium pyruvate	180 mg	Sigma, P4562	fresh
Stock D (100x)			prepare
$CaCl_2 \times 2\, H_2O$	1.26 g	Sigma,C4830	fresh
Stock E (10x)			prepare
HEPES solution	12.5 ml	Gibco,15630-056	fresh
(1 M, pH 7.2)			
Phenol red	2 ml	Sigma, P3532	
(50 mg/ml in water)			

Preparation of M2 and M16 media

For the preparation of 50 ml M2 and M16 media mix stocks A-E as indicated in the table below, fill up with water and add the BSA powder on the top of the solution. Gently shake the closed tube to allow the BSA to dissolve, filter sterilize and store in aliquots at 4°C up to two weeks. The pH value of M2 medium will be at 7.2 if equilibrated HEPES is used for preparation of stock E. If the pH value of M16 medium is not appropriate the medium can be filled in a cell culture flask (loose cap) and equilibrated in a 37°C incubator at 7.5% CO_2 for several hours.

Stock	M2	M16
A	5 ml	5 ml
B	0.8 ml	5 ml
C	0.5 ml	0.5 ml
D	0.5 ml	0.5 ml
E	4.2 ml	-
Water	39 ml	39 ml
BSA (Sigma, A3311)	200 mg	200 mg

Oil

For overlaying embryo cultures use 'embryo tested' mineral oil from Sigma No. M-8410) (or paraffin oil from Merck, No. 7161).

For injection system tubing we use a slightly more dense Paraffin oil (Merck, No. 7174).

Pronase

Dissolve protease (Sigma, P8811) from *Streptomyces griseus* (pronase) at a concentration of 0.5% in M2 medium. Centrifuge to remove insoluble material, filter sterilize and store in aliquots at - 20°C.

Tyrode's solution

Weigh the components into a 50 ml tube, fill up with water to 40 ml and adjust the pH to 2,5 with 5 M HCl using pH indicator strips. Fill up with water to 50 ml, filter sterilize and store in aliquots at -20°C.

Component	per 50 ml
NaCl	400 mg
KCl	10 mg
$CaCl_2$ x $2H_2O$	12 mg
$MgCl_2$ x $7H_2O$	5 mg

Glucose	50 mg
Polyvinylpyrrolidone	200 mg
(Sigma, P0930)	

APPENDIX B

Cre/loxP vectors

This section lists vectors which may help newcomers to easily perform Cre/loxP based gene targeting experiments. These vectors were constructed in K. Rajewsky's Lab in Cologne and are available for research purposes on request. For plasmid order forms, sequences and maps please refer to the web page: www.genetik.uni-koeln.de/gene_targeting/. For other Cre/loxP vectors see ref. (110). The neo gene expression vector pMC1NeopA (1) and FLP/FRT vectors (1, 72) are available from Stratagene. An HSV-tk expression vector has been previously described (19).

pGEMloxP

pGEMloxP was generated by designing oligonucleotide primers spanning the 34 bp loxP site and inserted into the polylinker of pGEM7Z (R. Pelanda, unpublished).

pGEM-30

pGEM-30 (58) contains a single loxP site which can be isolated by various restriction digests on a ~100 bp fragment. To test easily for the orientation of the ligated loxP site a larger fragment including the adjacent spacer fragment (LEU2 intron), which can be deleted in a subsequent cloning step, might be used.

pMMneoflox8

Plasmid pMMneoflox8 contains the neo gene expression cassette from pMC1NeopA (wildtype version, see ref. (82)) flanked by two loxP sites (M. Kraus, unpublished).

pL2neo

Plasmid pL2neo (61) contains the neo gene expression cassette from pMC1NeopA (wildtype version, see ref (82)) flanked by two loxP sites and an (irrelevant) intron of the yeast LEU2 gene. The floxed 1.4 kb cassette can be recovered by XbaI and SalI digestion. The insert from pL2neo, when included in a targeting vector, can be used for the removal of the neo gene from a targeted locus and the flox and delete strategy described in Chapters 5 and 6 (Section 1).

pIC-Cre and pMC-Cre

These plasmids contain the promoter/enhancer and pA regions from pMC1NeopA and were designed as Cre expression vectors for ES cells (58). The expression cassettes, but not the isolated Cre coding part, can be recovered by restriction digests (XhoI + SalI). pMC-Cre contains the SV40 large T antigen-derived nuclear localization signal N-terminal of the Cre sequence. Point mutations have been occasionally found in pMC-Cre.

Either pIC-Cre and pMC-Cre can be used for transient Cre expression to generate Cre-mediated deletions in ES cells (see Section 1).

pGH-1

pGH-1 (58) was constructed to generate and easily select for large deletions in ES cells (Chapter 7, Section 1). The neo gene (from pMC1NeopA) and HSV-tk gene (19) expression cassettes are flanked by a single loxP site.

pGK-Cre^{NLS}bpA

This plasmid (unpublished) can be used to recover the coding sequence of Cre (as a PstI/XbaI fragment) without promoter or pA regions for the construction of Cre expression vectors, e.g. for the production of transgenic mice. This plasmid can also be used for transient Cre expression to generate Cre-mediated deletions in ES cells (see Section 1). The N-terminal sequence includes a nuclear localization signal and is the same as in pMC-Cre (58).

pSVlacZT

pSVlacZT (F. Schwenk, unpublished) is a Cre recombination substrate vector which can be used to test Cre expression vectors by transient or stable transfection in cell lines. It is based on the β-galactosidase expression vector pCH110 (Pharmacia) which has been interrupted by the loxP flanked neo gene from pL2neo. β-Galactosidase is expressed from the SV40 promoter after Cre-mediated excision of the neo gene.

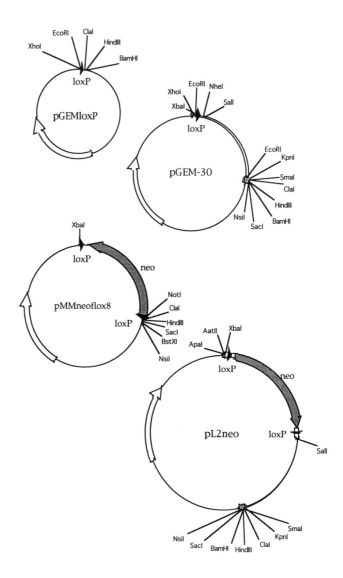

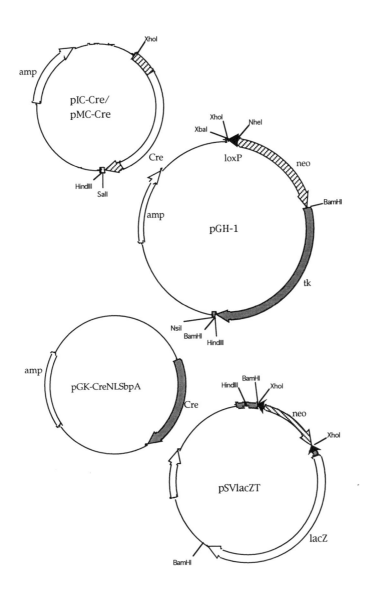

APPENDIX C

In situ detection of β-galactosidase activity

β-Galactosidase expression can be used to measure the efficiency of transient transfection of ES cells (e.g. using pCH110; Pharmacia) or to test Cre expression vectors cotransfected with the recombination substrate vector PSVlacZT (Appendix B). If ES cells are used, it is recommended to grow the cells first on EF cells and after transfection to plate them on gelatinized plates since (stained) individual ES cells can be more easily recognized on the latter plates. Count, however, only small (undifferentiated) stained ES cells as positive to exclude transfected feeder cells and differentiated ES cells from your record. Staining must be performed 2-3 days after transfection as plasmids are lost from the transfected cells with continuous cell divisions. For the Cre recombination assay it might be helpful in some cases to initially transfect the (neo gene containing) substrate vector PSVlacZT stably in your cell line and to test the Cre expression vector by transient transfection of stable transformants. In general, for a transfection and staining experiment you should include a control plate with nontransfected cells to assess for endogenous β-galactosidase activity in the cell line used.

Transient transfection is performed with the same conditions used for stable transfection (Chapter 15, Section 2) except that 25 - 50 μg of *supercoiled* plasmid DNA is used.

Staining protocol

1. If adherent cells (e.g ES cells) are used, wash culture plate(s) 1x with (MT)PBS. If cells are nonadherent spin ~10^4 cells on a slide using a cytocentrifuge.

2. Remove (MT)PBS and fix cells by addition of 4% formaldehyde in (MT)PBS (3 ml for 6 cm plate) and incubate 10 min at room temperature.

3. Wash plates 1x with (MT)PBS and add staining solution (2 ml for 6 cm plate).

4. Incubate 2 - 24 hours at 37°C in humid atmosphere and count the number of blue/nonstained cells. Staining develops rapidly (~2 hours) in a fraction of transfected cells but weak expressing cells can be detected only after prolonged incubation (~24 hours).

Staining solution:

5 mM $K3(Fe^{III}(CN)_6)$

5 mM $K4(Fe^{II}(CN)_6)$

2 mM $MgCl_2$

1 mg/ml X-Gal (5-bromo-chloro-3-indoyl-β–D-galactopyranosid)

Prepare fresh from the following stock solutions:

10 mM K3(Fe^{III}(CN)$_6$) (store light protected at 4°C)

10 mM K4(Fe^{II}(CN)$_6$) (store light protected at 4°C)

1 M MgCl$_2$

40 mg/ml X-Gal in DMSO (store at -20°C light protected)

For 1 volume of staining solution mix 1/2 volume of K3(Fe^{III}(CN)$_6$) and K4(Fe^{II}(CN)$_6$).stock solutions, add 1/500 volume of 1 M MgCl$_2$ and 1/40 volume of X-Gal stock solution.

APPENDIX D

Equipment suppliers

Capillaries:

•Embryo transfer needles: hard glass capillary tubes #32124, BDH Laboratory supplies, Poole BH 15 1TD, England.

•Injection & holder needles: borosilicate capillaries GC120-10, Clark Electromedical Instruments, Pangbourne, Reading RG8 7HU, England.

Cold light lamp

Schott KL 1500, Schott Glaswerke, Wiesbaden, Germany.

Darning Needle

Aggregation needle Cat. # DN-09 from Biological Lab Tool Maintainance and Service (BLS), H-1165 Budapest, Zsélyi Aladár u.31, Hungary.

Mycoplasma T.C. detection system

Gen-Probe, San Diego, CA 92121, USA.

Microforge

Microforge de Fonbrune, Bachofer Laboratoriumsgeräte, 72770 Reutlingen, Germany.

Micrometer syringes

AGLA Mikroliterspritze No. 12138, Vetter GmbH, 69168 Wiesloch-Baiertal, Germany.

Hamilton microliter syringes

100 μl: #710 or 1710, 250 μl: #725 or #1725, Hamilton, Vaduz, Switzerland.

Microscope/micromanipulator

Leitz Labovert with Leitz ground plate, 10x oculars, 10x, 16x, 32x, 40x phase contrast objectives for long working distance. Leitz micromanipulator Leica, D-6330 Wetzlar, Germany.

Needle puller

Kopf model 720, David Kopf instruments, Tujunga, CA 91042-0636, USA.

Oil for embryo cultures

Paraffin oil #8410, embryo tested, Sigma; or paraffin liquid #7161, Merck, Germany.

Oil for injection tubings

Paraffin highly liquid # 7174, Merck, Germany.

Stereomicroscope

Wild M5A, 15x oculars, Optik-Wild AG, Heerbrugg, Switzerland and/or
Nikon Tokyo, Japan.

Surgical instruments

Various instruments required for the animal surgery, Aesculap,
Tuttlingen, Germany.

Surgical silk

Perma-Hand Seide, 1.5 metric, 4-0, 100 m, No. EH 782P, Ethicon,
Norderstedt, Germany.

Wound clips and applier

Autoclip 9 mm No. 7631, Autoclip applier No. 7630, Clay Adams, Division
of Becton Dickinson & Co., Parsippany, NJ 07054, USA.

References

1.   Thomas, K.R., and M.R. Capecchi. 1987. Site-directed mutagenesis by gene targeting in mouse embryo-derived stem cells. *Cell* 51, no. 3:503-12.

2.   Brandon, E.P., R.L. Idzerda, and G.S. McKnight. 1995. Knockouts. Targeting the mouse genome: a compendium of knockouts (Part I). *Curr Biol* 5, no. 6:625-634.

3.   Brandon, E.P., R.L. Idzerda, and G.S. McKnight. 1995. Targeting the mouse genome: a compendium of knockouts (Part II). *Curr Biol* 5:758-765.

4.   Brandon, E.P., R.L. Idzerda, and G.S. McKnight. 1995. Targeting the mouse genome: a compendium of knockouts (Part III). *Curr Biol* 5, no. 8:873-881.

5.   Capecchi, M.R. 1989. The new mouse genetics: altering the genome by gene targeting. *Trends Genet* 5, no. 3:70-6.

6.   Koller, B.H., and O. Smithies. 1992. Altering genes in animals by gene targeting. *Ann Rev Immunol* 10:705-730.

7.   Smithies, O. 1993. Animal models of human genetic diseases. *Trends Genet* 9:112-116.

8.   Melton, D.W. 1994. Gene targeting in the mouse. *Bioessays* 16, no. 9:633-638.

9.   Hasty, P., R. Ramirez-Solis, R. Krumlauf, and A. Bradley. 1991. Introduction of a subtle mutation into the Hox-2.6 locus in embryonic stem cells [published erratum appears in *Nature* 1991 Sep 5;353(6339):94]. *Nature* 350, no. 6315:243-6.

10.   Rossant, J., and A. Nagy. 1995. Genome engineering: the new mouse genetics. *Nature Med* 1, no. 6:592-594.

11.   Rajewsky, K., H. Gu, R. Kühn, U.A.K. Betz, W. Müller, J. Roes, and F. Schwenk. 1996. Conditional gene targeting. *J Clin Invest* 98:600-603.

12.   Kühn, R., and F. Schwenk. 1997. Advances in gene targeting methods. *Curr Opin Immunol* in press.

13.   Marth, J.D. 1996. Recent advances in gene mutagenesis by site-directed recombination. *J Clin Invest* 97:1999-2002.

14.   Robertson, E.J. 1987. Teratocarcinomas and embryonic stem cells a practical approach. Practical Approach series, eds. D. Rickwood and B.D. Hames. IRL Press Limited, Oxford. 254 pp.

15.   Joyner, A.L., editor. 1993. *Gene Targeting A Practical Approach.* The Practical Approach Series. Edited by D. Rickwood and B.D. Hames. IRL Press at Oxford University Press, New York.

16.   Stewart, C.L. 1993. Production of chimeras between embryonic stem cells and embryos. *Methods Enzymol,* 225. 823-55 pp.

17.   Brinster, R.L., and M.R. Avarbock. 1994. Germline transmission of donor haplotype following spermatogonial transplantation [see comments]. *Proc Natl Acad Sci U S A* 91, no. 24:11303-7.

18.   Brinster, R.L., and J.W. Zimmermann. 1994. Spermatogenesis following male germ-cell transplantation [see comments]. *Proc Natl Acad Sci U S A* 91, no. 24:11298-302.

19.   Mansour, S.L., K.R. Thomas, and M.R. Capecchi. 1988. Disruption of the proto-oncogene int-2 in mouse embryo-derived stem cells: a general strategy for targeting mutations to non-selectable genes. *Nature* 336, no. 6197:348-52.

20.   Fiering, S., E. Epner, K. Robinson, Y. Zhuang, A. Telling, M. Hu, D.I.K. Martin, T. Enver, T.J. Ley, and M. Groundine. 1995. Targeted deletion of 5'HS2 of the murine beta-globin LCR reveals that it is not essential for proper regulation of the beta-globin locus. *Genes Dev.* 9, no. 18:2203-2213.

21.   Xu, Y., L. Davidson, F.W. Alt, and D. Baltimore. 1996. Deletion of the Ig kappa light chain intronic enhancer/matrix attachment region impairs but does not abolish V kappa J kappa rearrangement. *Immunity* 4, no. 4:377-385.

22.   Lerner, A., L. D'Adamio, A.C. Diener, L.K. Clayton, and E.L. Reinherz. 1993. CD3 zeta/eta/theta locus is colinear with and transcribed antisense to the gene encoding the transcription factor Oct-1. *J Immunol* 151, no. 6:3152-62.

23.   Koyasu, S., R.E. Hussey, L.K. Clayton, A. Lerner, R. Pedersen, P. Delany-Heiken, F. Chau, and E.L. Reinherz. 1994. Targeted disruption within the CD3 zeta/eta/phi/Oct-1 locus in mouse. *EMBO J* 13, no. 4:784-97.

24.   Ohno, H., S. Goto, S. Taki, T. Shirasawa, H. Nakano, S. Miyatake, T. Aoe, Y. Ishida, H. Maeda, T. Shirai, et al. 1994. Targeted disruption of the CD3 eta locus causes high lethality in mice: modulation of Oct-1 transcription on the opposite strand. *EMBO J* 13, no. 5:1157-65.

25.   Muller, U., N. Cristina, Z.W. Li, D.P. Wolfer, H.P. Lipp, T. Rulicke, S. Brandner, A. Aguzzi, and C. Weissmann. 1994. Behavioral and

anatomical deficits in mice homozygous for a modified beta-amyloid precursor protein gene. *Cell* 79, no. 5:755-765.

26.   Bradley, A., R. Ramirez-Solis, H. Zheng, P. Hasty, and A. Davis. 1992. Genetic manipulation of the mouse via gene targeting in embryonic stem cells. *Ciba Found Symp* 165:256-69; discussion 269-76.

27.   Hasty, P., and A. Bradley. 1993. Gene targeting vectors for mammalian cells. *In* Gene Targeting A practical approach. A.L. Joyner, editor. IRL Press at Oxford University Press, Oxford. 1-31.

28.   Valancius, V., and O. Smithies. 1991. Testing an "in-out" targeting procedure for making subtle genomic modifications in mouse embryonic stem cells. *Mol Cell Biol* 11, no. 3:1402-8.

29.   Stacey, A., A. Schnieke, J. McWhir, J. Cooper, A. Colman, and D.W. Melton. 1994. Use of double-replacement gene targeting to replace the murine alpha-lactalbumin gene with its human counterpart in embryonic stem cells and mice. *Mol Cell Biol* 14, no. 2:1009-16.

30.   Wu, H., X. Liu, and R. Jaenisch. 1994. Double replacement: strategy for efficient introduction of subtle mutations into the murine Col1a-1 gene by homologous recombination in embryonic stem cells. *Proc Natl Acad Sci U S A* 91, no. 7:2819-23.

31.   Askew, G.R., T. Doetschman, and J.B. Lingrel. 1993. Site-directed point mutations in embryonic stem cells: a gene-targeting tag-and-exchange strategy. *Mol Cell Biol* 13, no. 7:4115-24.

32.   Detloff, P.J., J. Lewis, S.W. John, W.R. Shehee, R. Langenbach, N. Maeda, and O. Smithies. 1994. Deletion and replacement of the mouse adult beta-globin genes by a "plug and socket" repeated targeting strategy. *Mol Cell Biol* 14, no. 10:6936-6943.

33.   Reid, L.H., E.G. Shesely, H.S. Kim, and O. Smithies. 1991. Cotransformation and gene targeting in mouse embryonic stem cells. *Mol Cell Biol* 11, no. 5:2769-77.

34.   Davis, A.C., M. Wims, and A. Bradley. 1992. Investigation of coelectroporation as a method for introducing small mutations into embryonic stem cells. *Mol Cell Biol* 12, no. 6:2769-76.

35.   Sternberg, N., and D. Hamilton. 1981. Bacteriophage P1 site-specific recombination. I. Recombination between loxP sites. *J Mol Biol* 150, no. 4:467-86.

36.   Hoess, R.H., M. Ziese, and N. Sternberg. 1982. P1 site-specific recombination: nucleotide sequence of the recombining sites. *Proc Natl Acad Sci U S A* 79, no. 11:3398-402.

37.   Hoess, R.H., and K. Abremski. 1984. Interaction of the bacteriophage P1 recombinase Cre with the recombining site loxP. *Proc Natl Acad Sci U S A* 81, no. 4:1026-9.

145

38.   Abremski, K., R. Hoess, and N. Sternberg. 1983. Studies on the properties of P1 site-specific recombination: evidence for topologically unlinked products following recombination. *Cell* 32, no. 4:1301-11.

39.   Hamilton, D.L., and K. Abremski. 1984. Site-specific recombination by the bacteriophage P1 lox-Cre system. Cre-mediated synapsis of two lox sites. *J Mol Biol* 178, no. 2:481-6.

40.   Sauer, B., and N. Henderson. 1990. Targeted insertion of exogenous DNA into the eukaryotic genome by the Cre recombinase. *New Biol* 2, no. 5:441-9.

41.   Hoess, R.H., A. Wierzbicki, and K. Abremski. 1986. The role of the loxP spacer region in P1 site-specific recombination. *Nucleic Acids Res* 14, no. 5:2287-300.

42.   Albert, H., Dale, E.C., Lee, E. and Ow, D.W. 1995. Site-specific integration of DNA into wild-type and mutant lox sites placed in the plant genome. *Plant J* 7:649-659.

43.   Hoess, R.H., and K. Abremski. 1985. Mechanism of strand cleavage and exchange in the Cre-lox site-specific recombination system. *J Mol Biol* 181, no. 3:351-62.

44.   Stark, W.M., M.R. Boocock, and D.J. Sherratt. 1992. Catalysis by site-specific recombinases [published erratum appears in *Trends Genet* 1993 Feb;9(2):45]. *Trends Genet* 8, no. 12:432-9.

45.   Mack, A., B. Sauer, K. Abremski, and R. Hoess. 1992. Stoichiometry of the Cre recombinase bound to the lox recombining site. *Nucleic Acids Res* 20, no. 17:4451-5.

46.   Hoess, R., A. Wierzbicki, and K. Abremski. 1985. Formation of small circular DNA molecules via an in vitro site-specific recombination system. *Gene* 40, no. 2-3:325-9.

47.   Austin, S., M. Ziese, and N. Sternberg. 1981. A novel role for site-specific recombination in maintenance of bacterial replicons. *Cell* 25, no. 3:729-36.

48.   Hochman, L., N. Segev, N. Sternberg, and G. Cohen. 1983. Site-specific recombinational circularization of bacteriophage P1 DNA. *Virology* 131, no. 1:11-7.

49.   Argos, P., A. Landy, K. Abremski, J.B. Egan, E. Haggard-Ljungquist, R.H. Hoess, M.L. Kahn, B. Kalionis, S.V. Narayana, L.d. Pierson, et al. 1986. The integrase family of site-specific recombinases: regional similarities and global diversity. *EMBO J* 5, no. 2:433-40.

50.   Sadowsky, P.D. 1993. Site-specific genetic recombination: hops, flips and flops. *FASEB J.* 7:760 - 767.

51.   Broach, J.R., and J.B. Hicks. 1980. Replication and recombination functions associated with the yeast plasmid, 2 mu circle. *Cell* 21, no. 2:501-8.

52.   Matsuzaki, H., H. Araki, and Y. Oshima. 1988. Gene conversion associated with site-specific recombination in yeast plasmid pSR1. *Mol Cell Biol* 8, no. 2:955-62.

53.   Abremski, K., and R. Hoess. 1984. Bacteriophage P1 site-specific recombination. Purification and properties of the Cre recombinase protein. *J Biol Chem* 259, no. 3:1509-14.

54.   Kilby, N.J., M.R. Snaith, and J.A. Murray. 1993. Site-specific recombinases: tools for genome engineering. *Trends Genet* 9, no. 12:413-21.

55.   Sauer, B., and N. Henderson. 1988. Site-specific DNA recombination in mammalian cells by the Cre recombinase of bacteriophage P1. *Proc Natl Acad Sci U S A* 85, no. 14:5166-70.

56.   Sauer, B., and N. Henderson. 1989. Cre-stimulated recombination at loxP-containing DNA sequences placed into the mammalian genome. *Nucleic Acids Res* 17, no. 1:147-61.

57.   Baubonis, W., and B. Sauer. 1993. Genomic targeting with purified Cre recombinase. *Nucleic Acids Res* 21, no. 9:2025-9.

58.   Gu, H., Y.R. Zou, and K. Rajewsky. 1993. Independent control of immunoglobulin switch recombination at individual switch regions evidenced through Cre-loxP-mediated gene targeting. *Cell* 73, no. 6:1155-64.

59.   Lakso, M., B. Sauer, B. Mosinger, Jr., E.J. Lee, R.W. Manning, S.H. Yu, K.L. Mulder, and H. Westphal. 1992. Targeted oncogene activation by site-specific recombination in transgenic mice. *Proc Natl Acad Sci U S A* 89, no. 14:6232-6.

60.   Orban, P.C., D. Chui, and J.D. Marth. 1992. Tissue- and site-specific DNA recombination in transgenic mice. *Proc Natl Acad Sci U S A* 89, no. 15:6861-5.

61.   Gu, H., J.D. Marth, P.C. Orban, H. Mossmann, and K. Rajewsky. 1994. Deletion of a DNA polymerase beta gene segment in T cells using cell type-specific gene targeting [see comments]. *Science* 265, no. 5168:103-6.

62.   Araki, K., M. Araki, J. Miyazaki, and P. Vassalli. 1995. Site-specific recombination of a transgene in fertilized eggs by transient expression of Cre recombinase. *Proc Natl Acad Sci U S A* 92, no. 1:160-164.

63.   Kühn, R., F. Schwenk, M. Aguet, and K. Rajewsky. 1995. Inducible gene targeting in mice. *Science* 269, no. 5229:1427-1429.

64.   Sauer, B. 1987. Functional expression of the cre-lox site-specific recombination system in the yeast Saccharomyces cerevisiae. *Mol Cell Biol* 7, no. 6:2087-96.

65.   Maruyama, I.N., and S. Brenner. 1992. A selective lambda phage cloning vector with automatic excision of the insert in a plasmid. *Gene* 120, no. 2:135-41.

66.   Dale, E.C., and D.W. Ow. 1990. Intra- and intermolecular site-specific recombination in plant cells mediated by bacteriophage P1 recombinase. *Gene* 91, no. 1:79-85.

67.   Bayley, C.C., M. Morgan, E.C. Dale, and D.W. Ow. 1992. Exchange of gene activity in transgenic plants catalyzed by the Cre-lox site-specific recombination system. *Plant Mol Biol* 18, no. 2:353-61.

68.   Qin, M., C. Bayley, T. Stockton, and D.W. Ow. 1994. Cre recombinase-mediated site-specific recombination between plant chromosomes. *Proc Natl Acad Sci U S A* 91, no. 5:1706-10.

69.   Jung, S., K. Rajewsky, and A. Radbruch. 1993. Shutdown of class switch recombination by deletion of a switch region control element. *Science* 259, no. 5097:984-7.

70.   Dymecki, S.M. 1996. Flp recombinase promotes site-specific DNA recombination in embryonic stem cells and transgenic mice. *Proc Natl Acad Sci U S A* 93:6191-6196.

71.   Buchholz, F., L. Ringrose, P.O. Angrand, F. Rossi, and A.F. Stewart. 1996. Different thermostabilities of FLP and Cre recombinases: Implications for applied site specific recombination. *Nucleic Acids Res.* 24:4256-4262.

72.   O'Gorman, S., D.T. Fox, and G.M. Wahl. 1991. Recombinase-mediated gene activation and site-specific integration in mammalian cells. *Science* 251, no. 4999:1351-5.

73.   Logie, C., and F. Stewart. 1995. Ligand-regulated site-specific recombination. *Proc. Natl. Acad. Sci. USA* 92:5940-5944.

74.   Golic, K.G., and S. Lindquist. 1989. The FLP recombinase of yeast catalyzes site-specific recombination in the Drosophila genome. *Cell* 59, no. 3:499-509.

75.   Golic, K.G. 1991. Site-specific recombination between homologous chromosomes in Drosophila. *Science* 252, no. 5008:958-61.

76.   Lyznik, L.A., L. Hirayama, K.V. Rao, A. Abad, and T.K. Hodges. 1995. Heat-inducible expression of Flp gene in maize cells. *Plant J.* 8, no. 2:177-186.

148

77.   Kilby, N.J., G.J. Davies, and M.R. Snaith. 1995. Flp recombinase in transgenic plants: constitutive activity in stably transformed tobacco and generation of marked cell clones in Arabidopsis. *Plant J.* 8, no. 5:637-652.

78.   te Riele, H., E.R. Maandag, and A. Berns. 1992. Highly efficient gene targeting in embryonic stem cells through homologous recombination with isogenic DNA constructs. *Proc Natl Acad Sci U S A* 89, no. 11:5128-32.

79.   Deng, C., and M.R. Capecchi. 1992. Reexamination of gene targeting frequency as a function of the extent of homology between the targeting vector and the target locus. *Mol Cell Biol* 12, no. 8:3365-71.

80.   Thomas, K.R., C. Deng, and M.R. Capecchi. 1992. High-fidelity gene targeting in embryonic stem cells by using sequence replacement vectors. *Mol Cell Biol* 12, no. 7:2919-23.

81.   Buchholz, F., P. Angrand, and A.F. Stewart. 1996. A simple assay to determine the functionality of Cre or Flp recombination targets in genomic manipulation constructs. *Nucleic Acids Research* 24:3118-3119.

82.   Yenofsky, R.L., M. Fine, and J.W. Pellow. 1990. A mutant neomycin phosphotransferase II gene reduces the resistance of transformants to antibiotic selection pressure. *Proc Natl Acad Sci U S A* 87, no. 9:3435-9.

83.   Soriano, P., C. Montgomery, R. Geske, and A. Bradley. 1991. Targeted disruption of the c-src proto-oncogene leads to osteopetrosis in mice. *Cell* 64, no. 4:693-702.

84.   Hasty, P., J. Rivera-Perez, and A. Bradley. 1992. The role and fate of DNA ends for homologous recombination in embryonic stem cells. *Mol Cell Biol* 12, no. 6:2464-74.

85.   Bollag, R.J., A.S. Waldman, and R.M. Liskay. 1989. Homologous recombination in mammalian cells. *Annu Rev Genet* 23:199-225.

86.   Hasty, P., J. Rivera-Perez, C. Chang, and A. Bradley. 1991. Target frequency and integration pattern for insertion and replacement vectors in embryonic stem cells. *Mol Cell Biol* 11, no. 9:4509-17.

87.   Pham, C.T.N., D.M. MacIvor, B.A. Hug, J.W. Heusel, and T.J. Ley. 1996. Long-range disruption of gene expression by a selectable marker cassette. *Proc Natl Acad Sci U S A* 93:13090-13095.

88.   Ferradini, L., H. Gu, A. De Smet, K. Rajewsky, C.A. Reynaud, and J.C. Weill. 1996. Rearrangement-enhancing element upstream of the mouse immunoglobulin kappa chain J cluster. *Science* 271, no. 5254:1416-1420.

89.   Taki, S., M. Meiering, and K. Rajewsky. 1993. Targeted insertion of a variable region gene into the immunoglobulin heavy chain locus [see comments]. *Science* 262, no. 5137:1268-71.

149

90.   Pelanda, R., S. Schall, R.M. Torres, and K. Rajewsky. 1996. A Prematurely Expressed Igk Transgene, but not a VkJk Gene Segment Targeted into the Igk Locus, Can Rescue B Cell Development in lambda-5-Deficient Mice. *Immunity* 5:229-239.

91.   Torres, R.M., H. Flaswinkel, M. Reth, and K. Rajewsky. 1996. Aberrant B cell development and immune response in mice with a compromised BCR complex. *Science* 272, no. 5269:1804-1808.

92.   Abuin, A., and A. Bradley. 1996. Recycling selectable markers in mouse embryonic stem cells. *Mol. Cell. Biol.* 16:1851-1856.

93.   Zhang, H., P. Hasty, and A. Bradley. 1994. Targeting frequency for deletion vectors in embryonic stem cells. *Mol Cell Biol* 14, no. 4:2404-10.

94.   Li, Z.-W., G. Stark, J. Götz, T. Rülicke, U. Müller, and C. Weissmann. 1996. Generation of mice with a 200-kb amyloid precursor protein gene deletion by Cre recombinase-mediated site-specific recombination in embryonic stem cells. *Proc Natl Acad Sci U S A* 93:6158-6162.

95.   Ramirez-Solis, R., P. Liu, and A. Bradley. 1995. Chromosome engineering in mice. *Nature* 378, no. 6558:720-724.

96.   Lakso, Alt, and Westphal. 1996. Efficient in vivo manipulation of mouse genomic sequences at the zygote stage. *Proc Natl Acad Sci U S A* 93:5860-5865.

97.   Schwenk, F., U. Baron, and K. Rajewsky. 1995. A cre-transgenic mouse strain for the ubiquitous deletion of loxP- flanked gene segments including deletion in germ cells. *Nucleic Acids Res* 23, no. 24:5080-1.

98.   Hanks, M., W. Wurst, L. Anson-Cartwhright, A. Auerbach, and A.L. Joyner. 1995. Rescue of the en-1 mutant phenotype by replacement of en-1 with en-2. *Science* 269:679-682.

99.   Zou, Y.R., W. Muller, H. Gu, and K. Rajewsky. 1994. Cre-loxP-mediated gene replacement: a mouse strain producing humanized antibodies. *Curr Biol* 4, no. 12:1099-1103.

100.   Smith, A.J.H., M.A. De Sousa, B. Kwabi-Addo, A. Heppell-Parton, H. Impey, and P. Rabbits. 1995. A site-directed chromosomal translocation induced in embryonic stem cells by Cre-*lox*P recombination. *Nature Genet* 9:376-385.

101.   Van Deursen, J., M. Fornerod, B. Van Rees, and G. Grosveld. 1995. Cre-mediated site-specific translocation between nonhomologous mouse chromosomes. *Proc Natl Acad Sci U S A* 92, no. 16:7376-7380.

102.   Ow, D.W. 1996. Recombinase-directed chromosome engineering in plants. *Curr Opin Biotechnol* 7:181-186.

150

103.	Fukushige, S., and B. Sauer. 1992. Genomic targeting with a positive-selection lox integration vector allows highly reproducible gene expression in mammalian cells. *Proc Natl Acad Sci U S A* 89, no. 17:7905-9.

104.	DiSanto, J.P., W. Müller, D. Guy-Grand, A. Fischer, and K. Rajewsky. 1995. Lymphoid development in mice with a targeted deletion of the interleukin-2 receptor gamma chain. *Proc Natl Acad Sci U S A* 92:377-381.

105.	Hennet, T., F.K. Hagen, L.A. Tabak, and J.D. Marth. 1995. T-cell-specific deletion of a polypeptide N-acetylgalactosaminyl- transferase gene by site-directed recombination. *Proc Natl Acad Sci U S A* 92, no. 26:12070-12074.

106.	Tsien, J.Z., P.T. Huerta, and S. Tonegawa. 1996. The essential role of hippocampal CA1 NMDA receptor-dependent synaptic plasticity in spatial memory. *Cell* 87:1327 - 1338.

107.	Ashfield, R., A.J. Patel, S.A. Bossone, H. Brown, R.D. Campbell, K.B. Marcu, and N.J. Proudfoot. 1994. MAZ-dependent termination between closely spaced human complement genes. *EMBO J* 13, no. 23:5656-5667.

108.	Rohlmann, A., M. Gotthardt, T.E. Willnow, R.E. Hammer, and J. Herz. 1996. Sustained somatic gene inactivation by viral transfer of Cre recombinase. *Nature Biotechnol* 14:1562-1565.

109.	Wang, Y., L.A. Krushel, and G.M. Edelman. 1996. Targeted DNA recombination in vivo using an adenovirus carrying the cre recombinase gene. *Proc Natl Acad Sci U S A* 93:3932-3936.

110.	Sauer, B. 1993. Manipulation of transgenes by site-specific recombination: use of Cre recombinase. *In* Guide to Techniques in Mouse Development, vol. 225. P.M. Wassarman and M.L. DePamphilis, editors. Academic Press, Inc., San Diego. 890-900.

111.	Sauer, B. 1992. Identification of cryptic lox sites in the yeast genome by selection for Cre-mediated chromosome translocations that confer multiple drug resistance. *J Mol Biol* 223, no. 4:911-28.

112.	Gordon, J.W. 1995. Production of Transgenic Mice. *In* Guide to Techniques in Mouse Development, vol. 225. P.M. Wassarman and M.L. DePamphilis, editors. Academic Press, Inc., San Diego. 1021.

113.	Hogan, B., R. Beddington, F. Costantini, and E. Lacy. 1994. Manipulating the Mouse Embryo. Cold Spring Harbor Laboratory Press, Plainview, New York. 497 pp.

114.	Bronson, S.K., E.G. Plaehn, K.D. Kluckman, J.R. Hagaman, N. Maeda, and O. Smithies. 1996. Single-copy transgenic mice with chosen-site integration. *Proc Natl Acad Sci U S A* 93:9067-9072.

115.   Jasin, M., M.E. Moynahan, and C. Richardson. 1996. Targeted transgenesis. *Proc Natl Acad Sci U S A* 93:8804-8808.

116.   Palmiter, R.D., and R.L. Brinster. 1986. Germ-line transformation of mice. *Annu Rev Genet* 20:465-99.

117.   Schwenk, F., B. Sauer, N. Kukoc, R. Hoess, W. Müller, C. Kocks, R. Kühn, and K. Rajewsky. Generation of Cre recombinase-specific monoclonal antibodies to characterize the pattern of Cre expression in *cre*-transgenic mouse strains. *submitted*.

118.   Brinster, R.L., J.M. Allen, R.R. Behringer, R.E. Gelinas, and R.D. Palmiter. 1988. Introns increase transcriptional efficiency in transgenic mice. *Proc Natl Acad Sci U S A* 85, no. 3:836-40.

119.   Chaffin, K.E., C.R. Beals, T.M. Wilkie, K.A. Forbush, M.I. Simon, and R.M. Perlmutter. 1990. Dissection of thymocyte signaling pathways by in vivo expression of pertussis toxin ADP-ribosyltransferase. *EMBO J* 9, no. 12:3821-9.

120.   Gossen, M., S. Freundlieb, G. Bender, G. Muller, W. Hillen, and H. Bujard. 1995. Transcriptional activation by tetracyclines in mammalian cells. *Science* 268, no. 5218:1766-1769.

121.   No, D., T.P. Yao, and R.M. Evans. 1996. Ecdysone-inducible gene expression in mammalian cells and transgenic mice. *Proc Natl Acad Sci U S A* 93:3346-3351.

122.   St-Onge, L., P.A. Furth, and P. Gruss. 1996. Temporal control of the Cre recombinase in transgenic mice by a tetracycline responsive promoter. *Nucleic Acids Res* 24:3875 - 3877.

123.   Furth, P.A., L. St Onge, H. Boger, P. Gruss, M. Gossen, A. Kistner, H. Bujard, and L. Hennighausen. 1994. Temporal control of gene expression in transgenic mice by a tetracycline-responsive promoter. *Proc Natl Acad Sci U S A* 91, no. 20:9302-9306.

124.   Kistner, A., M. Gossen, F. Zimmermann, J. Jerecic, C. Ullmer, H. Lübbert, and H. Bujard. 1996. Doxycycline-mediated quantitative and tissue-specific control of gene expression in transgenic mice. *Proc Natl Acad Sci U S A* 93:10933-10938.

125.   Schultze, N., Y. Burki, Y. Lang, U. Certa, and H. Bluethmann. 1996. Efficient control of gene expression by single step integration of the tetracycline system in transgenic mice. *Nature Biotechnology* 14:499 - 503.

126.   Picard, D. 1993. Steroid-binding domains for regulating the functions of heterologous proteins in cis. *Trends Cell Biol* 3:278 - 280.

127.   Zhang, Y., C. Riesterer, A.M. Ayrall, F. Sablitzky, T.D. Littlewood, and M. Reth. 1996. Inducible site-directed recombination in mouse embryonic stem cells. *Nucleic Acids Res* 24, no. 4:543-548.

128. Feil, R., J. Brocard, B. Mascrez, M. LeMeur, D. Metzger, and P. Chambon. 1996. Ligand-activated site-specific recombination in mice. *Proc Natl Acad Sci U S A* 93:10887-10890.

129. Kellendonk, C., F. Tronche, A.P. Monaghan, P.O. Angrand, F. Stewart, and G. Schutz. 1996. Regulation of Cre recombinase activity by the synthetic steroid RU 486. *Nucleic Acids Res* 24, no. 8:1404-1411.

130. Rickert, R.C., J. Roes, and K. Rajewsky. 1997. B lymphocyte-specific, Cre-mediated mutagenesis in mice. *Nucleic Acids Res* 25, no. 6 1317-1318.

131. Rickert, R.C., K. Rajewsky, and J. Roes. 1995. Impairment of T cell-dependent B-cell responses and B-1 cell development in CD19 deficient mice. *Nature* 376:352-355.

132. Tsien, J.Z., D.F. Chen, D. Gerber, C. Tom, E.H. Mercer, D.J. Anderson, M. Mayford, E.R. Kandel, and S. Tonegawa. 1996. Subregion- and cell type-restricted gene knockout in mouse brain. *Cell* 87:1317-1326.

133. Betz, U.A.K., C.A.J. Voßhenrich, K. Rajewsky, and W. Müller. 1996. Bypass of lethality with mosaic mice generated by Cre-*loxP*-mediated recombination. *Curr Biol* 6:1307-1316.

134. Silver, L.M. 1995. Mouse Genetics, Concepts and Applications. Oxford University Press, New York. 362 pp.

135. Müller, W., R. Kühn, and K. Rajewsky. 1991. Major histocompatibility complex class II hyperexpression on B cells in interleukin 4-transgenic mice does not lead to B cell proliferation and hypergammaglobulinemia. *Eur J Immunol* 21, no. 4:921-5.

136. Martin, G.R. 1981. Isolation of a pluripotent cell line from early mouse embryos cultured in medium conditioned by teratocarcinoma stem cells. *Proc Natl Acad Sci U S A* 78, no. 12:7634-8.

137. Evans, M.J., and M.H. Kaufman. 1981. Establishment in culture of pluripotential cells from mouse embryos. *Nature* 292, no. 5819:154-6.

138. Hooper, M.L. 1992. Embryonal Stem Cells. Modern Genetics, ed. H.J. Evans, 1. Harwood Academics Publishers GmbH, Switzerland. 147 pp.

139. Martin, G.R. 1980. Teratocarcinomas and mammalian embryogenesis. *Science* 209, no. 4458:768-76.

140. Bradley, A., M. Evans, M.H. Kaufman, and E. Robertson. 1984. Formation of germ-line chimaeras from embryo-derived teratocarcinoma cell lines. *Nature* 309, no. 5965:255-6.

141. Gossler, A., T. Doetschman, R. Korn, E. Serfling, and R. Kemler. 1986. Transgenesis by means of blastocyst-derived embryonic stem cell lines. *Proc Natl Acad Sci U S A* 83, no. 23:9065-9.

153

142.   Robertson, E., A. Bradley, M. Kuehn, and M. Evans. 1986. Germ-line transmission of genes introduced into cultured pluripotential cells by retroviral vector. *Nature* 323, no. 6087:445-8.

143.   Doetschman, T.C., H. Eistetter, M. Katz, W. Schmidt, and R. Kemler. 1985. The in vitro development of blastocyst-derived embryonic stem cell lines: formation of visceral yolk sac, blood islands, and myocardium. *J Embryol Exp Morph* 87:27-45.

144.   Hooper, M., K. Hardy, A. Handyside, S. Hunter, and M. Monk. 1987. HPRT-deficient (Lesch-Nyhan) mouse embryos derived from germline colonization by cultured cells. *Nature* 326:292.

145.   Handyside, A.H., G.T. O'Neill, M. Jones, and M. Hooper. 1989. Use of BRL-conditioned medium in combination with feeder layers to isolate a diploid embryonal stem cell line. *Roux Arch Dev Biol* 198:48-56.

146.   McMahon, A.P., and A. Bradley. 1990. The Wnt-1 (int-1) proto-oncogene is required for development of a large region of the mouse brain. *Cell* 62, no. 6:1073-85.

147.   Nagy, A., J. Rossant, R. Nagy, W. Abramow-Newerly, and J.C. Roder. 1993. Derivation of completely cell culture-derived mice from early-passage embryonic stem cells. *Proc Natl Acad Sci U S A* 90, no. 18:8424-8.

148.   Li, E., T.H. Bestor, and R. Jaenisch. 1992. Targeted mutation of the DNA methyltransferase gene results in embryonic lethality. *Cell* 69, no. 6:915-26.

149.   Magin, T.M., J. McWhir, and D.W. Melton. 1992. A new mouse embryonic stem cell line with good germ line contribution and gene targeting frequency. *Nucleic Acids Res* 20, no. 14:3795-6.

150.   Ledermann, B., and K. Burki. 1991. Establishment of a germ-line competent C57BL/6 embryonic stem cell line. *Exp Cell Res* 197, no. 2:254-8.

151.   Abbondanzo, S.J., I. Gadi, and C.L. Stewart. 1993. Derivation of embryonic stem cell lines. *Methods Enzymol*, 225. 803-23 pp.

152.   Williams, R.L., D.J. Hilton, S. Pease, T.A. Willson, C.L. Stewart, D.P. Gearing, E.F. Wagner, D. Metcalf, N.A. Nicola, and N.M. Gough. 1988. Myeloid leukaemia inhibitory factor maintains the developmental potential of embryonic stem cells. *Nature* 336, no. 6200:684-7.

153.   Smith, A.G., J.K. Heath, D.D. Donaldson, G.G. Wong, J. Moreau, M. Stahl, and D. Rogers. 1988. Inhibition of pluripotential embryonic stem cell differentiation by purified polypeptides. *Nature* 336, no. 6200:688-90.

154.   Rottem, S., and M.F. Barile. 1993. Beware of mycoplasmas. *Trends Biochem Sci* 11:143-151.

155.   Kühn, R., K. Rajewsky, and W. Müller. 1991. Generation and analysis of Interleukin-4 deficient mice. *Science* 254:707-710.

156.   Riethmacher, D., V. Brinkmann, and C. Birchmeier. 1995. A targeted mutation in the mouse E-cadherin gene results in defective preimplantation development. *Proc Natl Acad Sci U S A* 92, no. 3:855-859.

157.   Tanaka, T., S. Akira, K. Yoshida, M. Umemoto, Y. Yoneda, N. Shirafuji, H. Fujiwara, S. Suematsu, N. Yoshida, and T. Kishimoto. 1995. Targeted disruption of the NF-IL6 gene discloses its essential role in bacteria killing and tumor cytotoxicity by macrophages. *Cell* 80, no. 2:353-361.

158.   Nishinakamura, R., N. Nakayama, Y. Hirabayashi, T. Inoue, D. Aud, T. NcNeil, S. Azuma, S. Yoshida, Y. Toyoda, K. Arai, A. Miyajama, and R. Murray. 1995. Mice deficient for the IL-3/GM-CSF/IL-5 βc receptor exhibit lung pathology and impaired immune response while β_{IL3} receptor deficient mice are normal. *Immunity* 2:211-222.

159.   Silvers, W.K. 1979. The Coat Colors of Mice. Springer-Verlag, New York.

160.   Bronson, S.K., O. Smithies, and J.T. Mascarello. 1995. High incidence of XXY and XYY males among the offspring of female chimeras from embryonic stem cells. *Proc Natl Acad Sci U S A* 92:3120-3123.

161.   Wong, E.A., and M.R. Capecchi. 1987. Homologous recombination between coinjected DNA sequences peaks in early to mid-S phase. *Mol Cell Biol* 7, no. 6:2294-5.

162.   Kim, H.-S., and O. Smithies. 1988. Recombinant fragment assay for gene targetting based on the polymerase chain reaction. *Nucleic Acids Res* 16, no. 18:8887-9003.

163.   Laird, P.W., A. Zijderveld, K. Linders, M.A. Rudnicki, R. Jaenisch, and A. Berns. 1991. Simplified mammalian DNA isolation procedure. *Nucleic Acids Res* 19, no. 15:4293.

164.   Chen, J., R. Lansford, V. Stewart, F. Young, and F.W. Alt. 1993. RAG-2-deficient blastocyst complementation: an assay of gene function in lymphocyte development. *Proc Natl Acad Sci U S A* 90, no. 10:4528-32.

165.   Nagy, A., and J. Rossant. 1996. Targeted mutagenesis: analysis of phenotype without germ line transmission. *J Clin Invest* 97, no. 6:1360-1365.

166.   Nagy, A., E. Gocza, E.M. Diaz, V.R. Prideaux, E. Ivanyi, M. Markkula, and J. Rossant. 1990. Embryonic stem cells alone are able to support fetal development in the mouse. *Development* 110, no. 3:815-21.

167.   Carmeliet, P., V. Ferreira, G. Breier, S. Pollefeyt, L. Kieckens, M. Gertsenstein, M. Fahrig, A. Vandenhoeck, K. Harpal, C. Eberhardt, C. Declercq, J. Pawling, L. Moons, D. Collen, W. Risau, and A. Nagy. 1996.

155

Abnormal blood vessel development and lethality in embryos lacking a single VEGF allele. *Nature* 380, no. 6573:435-439.

168.   Wood, S.A., W.S. Pascoe, C. Schmidt, R. Kemler, M.J. Evans, and N.D. Allen. 1993. Simple and efficient production of embryonic stem cell-embryo chimeras by coculture. *Proc Natl Acad Sci U S A* 90, no. 10:4582-5.

169.   Wood, S.A., N.D. Allen, J. Rossant, A. Auerbach, and A. Nagy. 1993. Non-injection methods for the production of embryonic stem cell-embryo chimaeras. *Nature* 365, no. 6441:87-9.

170.   Nagy, A., and J. Rossant. 1993. Production of completely ES cell-derived fetuses. *In* Gene targeting: a practical approach. A.L. Joyner, editor. Oxford University Press, Oxford.

171.   Rugh, R. 1990. The mouse: its reproduction and development. Oxford Science Publications, Oxford.

172.   Green, E.L. 1966. Biology of the Laboratory Mouse. 2nd ed. McGraw-Hill, New York.

173.   Kovacs, M.S., L. Lowe, and M.R. Kuehn. 1993. Use of superovulated mice as embryo donors for ES cell injection chimaeras. *Lab Animal Sci* 43:91-93.

174.   Flecknell, P.A. 1993. Anesthesia and perioperative care. *Methods Enzymol* 225:16-33.

175.   Garcia, J., S.D. Kholkute, and W.R. Dukelow. 1993. The efficiacy of stored pregnant mares' serum gonadotropin and human chorionic gonadotropin on inducing ovulation in mice. *Lab. Animal Sci.* 43:198-199.

Selected References for Further Reading

ES cells

Robertson, E.J. 1987. Teratocarcinomas and embryonic stem cells a practical approach. Practical Approach series, eds. D. Rickwood and B.D. Hames. IRL Press Limited, Oxford. 254 pp.

Hooper, M.L. 1992. Embryonal Stem Cells. Modern Genetics, ed. H.J. Evans, 1. Harwood Academics Publishers GmbH, Switzerland. 147 pp.

Gene targeting

Joyner, A.L., editor. 1993. *Gene Targeting A Practical Approach*. The Practical Approach Series. Edited by D. Rickwood and B.D. Hames. IRL Press at Oxford University Press, New York.

Experimental Embryology

Hogan, B., R. Beddington, F. Costantini, and E. Lacy. 1994. Manipulating the Mouse Embryo. Cold Spring Harbor Laboratory Press, Plainview, New York. 497 pp.

Wassarman, P.M., and M.L. DePamphilis. 1993. Guide to Techniques in Mouse Development. Methods in Enzymology, 225. Academic Press, San Diego.

Joyner, A.L., editor. 1993. *Gene Targeting A Practical Approach*. The Practical Approach Series. Edited by D. Rickwood and B.D. Hames. IRL Press at Oxford University Press, New York.

Kaufman, M.H. 1992. An atlas of Mouse Development. Academic Press, San Diego.

Genetics and mouse strains

Silver, L.M. 1995. Mouse Genetics, Concepts and Applications. Oxford University Press, New York. 362 pp.

Lyon, M.F., and A.G. Searle. 1989. Genetic Variants of and Strains of the Laboratory Mouse. 2nd ed. Oxford University Press, Oxford.

General Biology

Green, E.L. 1966. Biology of the Laboratory Mouse. 2nd ed. McGraw-Hill, New York.

Silvers, W.K. 1979. The Coat Colors of Mice. Springer-Verlag, New York.

Disease and Welfare

Wolfensohn, S., and M. Lloyd. 1994. Handbook of Laboratory Animal Management and Welfare. Oxford University Press, Oxford.

Foster, H.L., J.D. Small, and J.G. Fox. 1982. The mouse in Biomedical Research, Volume II. Academic Press, New York.

Commitee in Infectious Diseases of Mice and Rats. 1991. Infectious Diseases of Mice and Rats. National Academy Press, Washington, D.C.

Useful www pages

The Jackson Laboratory Home page:

http://www.jax.org

TBASE - Transgenic / Targeted Mutation database:

http://www.gdb.org/Dan/tbase/tbase.html

Mouse and Rat Research Home Page:

http://www.cco.caltech.edu/~mercer/htmls/rodent_page.html

Frontiers in Bioscience:

http://bioinformatics.weizmann.ac.il/bioscience/knockout/

knochome.htm

Index